INTRODUCTORY ANALYSIS

INTRODUCTORY ANALYSIS

A Deeper View of Calculus

Richard J. Bagby

Department of Mathematical Sciences
New Mexico State University
Las Cruces, New Mexico

HARCOURT
ACADEMIC
PRESS

San Diego San Francisco New York Boston London Toronto Sydney Tokyo

Sponsoring Editor	Barbara Holland
Production Editor	Julie Bolduc
Editorial Coordinator	Karen Frost
Marketing Manager	Marianne Rutter
Cover Design	Richard Hannus, Hannus Design Associates
Copyeditor	Amy Mayfield
Composition	TeXnology, Inc./MacroTEX
Printer	Maple-Vail Book Manufacturing Group

This book is printed on acid-free paper. ∞

ACADEMIC PRESS
A Harcourt Science and Technology Company
525 B Street, Suite 1900, San Diego, CA 92101-4495, USA
http://www.academicpress.com

Academic Press
Harcourt Place, 32 Jamestown Road, London NW1 7BY, UK

Harcourt/Academic Press
200 Wheeler Road, Burlington, MA 01803
http://www.harcourt-ap.com

Library of Congress Catalog Card Number: 00-103265
International Standard Book Number: 0-12-072550-9

Printed in the United States of America
00 01 02 03 04   MB   9 8 7 6 5 4 3 2 1

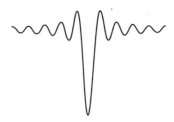

CONTENTS

III

LIMITS

IV

THE DERIVATIVE

V

THE RIEMANN INTEGRAL

VI

EXPONENTIAL AND LOGARITHMIC FUNCTIONS

VII

CURVES AND ARC LENGTH

VIII

SEQUENCES AND SERIES OF FUNCTIONS

IX

ADDITIONAL COMPUTATIONAL METHODS

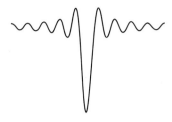

ACKNOWLEDGMENTS

would like to thank many persons for the support and assistance that I have received while writing this book. Without the support of my department I might never have begun, and the feedback I have received from my students and from reviewers has been invaluable. I would especially like to thank Professors William Beckner of University of Texas at Austin, Jung H. Tsai of SUNY at Geneseo and Charles Waters of Mankato State University for their useful comments. Most of all I would like to thank my wife, Susan; she has provided both encouragement and important technical assistance.

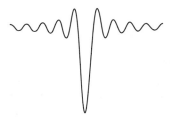

PREFACE

I ntroductory real analysis can be an exciting course; it is the gateway to an impressive panorama of higher mathematics. But for all too many students, the excitement takes the form of anxiety or even terror; they are overwhelmed. For many, their study of mathematics ends one course sooner than they expected, and for many others, the doorways that should have been opened now seem rigidly barred. It shouldn't have to be that way, and this book is offered as a remedy.

GOALS FOR *INTRODUCTORY ANALYSIS*

The goals of first courses in real analysis are often too ambitious. Students are expected to solidify their understanding of calculus, adopt an abstract point of view that generalizes most of the concepts, recognize how explicit examples fit into the general theory and determine whether they satisfy appropriate hypotheses, and not only learn definitions, theorems, and proofs but also learn how to construct valid proofs and relevant examples to demonstrate the need for the hypotheses. Abstract properties such as countability, compactness and connectedness must be mastered. The students who are up to such a challenge emerge ready to take on the world of mathematics.

A large number of students in these courses have much more modest immediate needs. Many are only interested in learning enough mathematics to be a good high-school teacher instead of to prepare for high-level mathematics. Others seek an increased level of mathematical maturity, but something less than a quantum leap is desired. What they need is a new understanding of calculus as a mathematical theory — how to study it in terms of assumptions and consequences, and then check whether the needed assumptions are actually satisfied in specific cases. Without such an understanding, calculus and real analysis seem almost unrelated in spite of the vocabulary they share, and this is why so many good calculus students are overwhelmed by the demands of higher mathematics. Calculus students come to expect regularity but analysis students must learn to expect irregularity; real analysis sometimes shows that incomprehensible levels of pathology are not only possible but theoretically ubiquitous. In calculus courses, students spend most of their energy using finite procedures to find solutions, while analysis addresses questions of existence when there may not even be a finite algorithm for recognizing a solution, let alone for producing one. The obstacle to studying mathematics at the next level isn't just the inherent difficulty of learning definitions, theorems, and proofs; it is often the lack of an adequate model for interpreting the abstract concepts involved. This is why most students need a different understanding of calculus before taking on the abstract ideas of real analysis. For some students, such as prospective high-school teachers, the next step in mathematical maturity may not even be necessary.

The book is written with the future teacher of calculus in mind, but it is also designed to serve as a bridge between a traditional calculus sequence and later courses in real or numerical analysis. It provides a view of calculus that is now missing from calculus books, and isn't likely to appear any time soon. It deals with derivations and justifications instead of calculations and illustrations, with examples showing the need for hypotheses as well as cases in which they are satisfied. Definitions of basic concepts are emphasized heavily, so that the classical theorems of calculus emerge as logical consequences of the definitions, and not just as reasonable assertions based on observations. The goal is to make this knowledge accessible without diluting it. The approach is to provide clear and complete explanations of the fundamental concepts, avoiding topics that don't contribute to reaching our objectives.

APPROACH

To keep the treatments brief yet comprehensible, familiar arguments have been re-examined, and a surprisingly large number of the traditional concepts of analysis have proved to be less than essential. For example, open and closed intervals are needed but open and closed sets are not, sequences are needed but subsequences are not, and limits are needed but methods for finding limits are not. Another key to simplifying the development is to start from an appropriate level. Not surprisingly, completeness of the real numbers is introduced as an axiom instead of a theorem, but the axiom takes the form of the nested interval principle instead of the existence of suprema or limits. This approach brings the power of sequences and their limits into play without the need for a fine understanding of the difference between convergence and divergence. Suprema and infima become more understandable, because the proof of their existence explains what their definition really means. By emphasizing the definition of continuity instead of limits of sequences, we obtain remarkably simple derivations of the fundamental properties of functions that are continuous on a closed interval:

- existence of intermediate values

- existence of extreme values

- uniform continuity.

Moreover, these fundamental results come early enough that there is plenty of time to develop their consequences, such as the mean value theorem, the inverse function theorem, and the Riemann integrability of continuous functions, and then make use of these ideas to study the elementary transcendental functions. At this stage we can begin mainstream real analysis topics: continuity, derivatives, and integrals of functions defined by sequences and series.

The coverage of the topics studied is designed to explain the concepts, not just to prove the theorems efficiently. As definitions are given they are explained, and when they seem unduly complicated the need for the complexity is explained. Instead of the definition - theorem - proof format often used in sophisticated mathematical expositions, we try to see how the definitions evolve to make further developments possible. The rigor is present, but the formality is avoided as much as possible. In general, proofs are given in their entirety rather in outline form; the reader isn't left with a sequence of exercises to complete them.

Exercises at the end of each section are designed to provide greater familiarity with the topics treated. Some clarify the arguments used in the text by having the reader develop parallel ones. Others ask the reader to determine how simple examples fit into the general theory, or give examples that highlight the relevance of various conditions. Still others address peripheral topics that the reader might find interesting, but that were not necessary for the development of the basic theory. Generally the exercises are not repetitive; the intent is not to provide practice for working exercises of any particular type, and so there are few worked examples to follow. Computational skill is usually important in calculus courses but that is not the issue here; the skills to be learned are more in the nature of making appropriate assumptions and working out their consequences, and determining whether various conditions are satisfied. Such skills are much harder to develop, but well worth the effort. They make it possible to do mathematics.

ORGANIZATION AND COVERAGE

The first seven chapters treat the fundamental concepts of calculus in a rigorous manner; they form a solid core for a one-semester course. The first chapter introduces the concepts we need for working in the real number system, and the second develops the remarkable properties of continuous functions that make a rigorous development of calculus possible. Chapter 3 is a deliberately brief introduction to limits, so that the fundamentals of differentiation and integration can be reached as quickly as possible. It shows little more than how continuity allows us to work with quantities given as limits. The fourth chapter studies differentiability; it includes a development of the implicit function theorem, a result that is not often presented at this level. Chapter 5 develops the theory of the Riemann integral, establishing the equivalence of Riemann's definition with more convenient ones and treating the fundamental theorem even when the integrand fails to be a derivative. The sixth chapter studies logarithms and exponents from an axiomatic point of view that leads naturally to formulas for them, and the seventh studies arc length geometrically before examining the connections between arc length and calculus.

Building on this foundation, Chapter 8 gets into mainstream real analysis, with a deeper treatment of limits so that we can work with sequences and series of functions and investigate questions of continuity, differentiability, and integrability. It includes the construction of a function that is continuous everywhere but nowhere differentiable or monotonic, showing that calculus deals with functions much more complicated than we can visualize, and the theory of power series is developed far enough to prove that each convergent power series is the Taylor series for its sum. The final chapter gives a careful analysis of some additional topics that are commonly learned in calculus but rarely explained fully. They include L'Hôpital's rule and an analysis of the error in Simpson's rule and Newton's method; these could logically have been studied earlier but were postponed because they were not needed for further developments. They could serve as independent study projects at various times in a course, rather than studied at the end.

A few historical notes are included, simply because they are interesting. While a historical understanding of calculus is also desirable, some traditional calculus texts, such as the one by Simmons [3], already meet this need.

GETTING THE MOST FROM THIS BOOK

Books should be read and mathematics should be done; students should expect to do mathematics while reading this book. One of my primary goals was to make it read easily, but reading it will still take work; a smooth phrase may be describing a difficult concept. Take special care in learning definitions; later developments will almost always require a precise understanding of just exactly what they say. Be especially wary of unfamiliar definitions of familiar concepts; that signals the need to adopt an unfamiliar point of view, and the key to understanding much of mathematics is to examine it from the right perspective. The definitions are sometimes more complex than they appear to be, and understanding the stated conditions may involve working through several logical relationships. Each reader should try to think of examples of things that satisfy the relevant conditions and also try to find examples of things that don't; understanding how a condition can fail is a key part of understanding what it really means.

Take the same sort of care in reading the statement of a theorem; the hypotheses and the conclusion need to be identified and then understood. Instead of reading a proof passively, the reader should work through the steps described and keep track of what still needs to be done, question why the approach was taken, check the logic, and look for potential pitfalls. A writer of mathematics usually expects this level of involvement, and that's why the word "we" appears so often in work by a single author. With an involved reader, the mathematics author can reveal the structure of an argument in a way that is much more enlightening than an overly detailed presentation would be.

Pay close attention to the role of stated assumptions. Are they made simply for the purposes of investigation, to make exploration easier, or are they part of a rigorous argument? Are the assumptions known to be true whenever the stated hypotheses are satisfied, or do they simply correspond to special cases being considered separately? Or is an assumption made solely for the sake of argument, in order to show that it can't be true?

Mastering the material in this book will involve doing mathematics actively, and the same is probably true of any activity that leads to greater mathematical knowledge. It is work, but it is rewarding work, and it can be enjoyable.

I

THE REAL NUMBER SYSTEM

From counting to calculus, the methods we use in mathematics are intimately related to the properties of the underlying number system. So we begin our study of calculus with an examination of real numbers and how we work with them.

I FAMILIAR NUMBER SYSTEMS

The first numbers we learn about are the natural numbers N, which are just the entire collection of positive integers 1, 2, 3, ... that we use for counting. But the natural number system is more than just a collection of numbers. It has additional structure, and the elements of N can be identified by their role in this structure as well as by the numerals we ordinarily use. For example, each natural number $n \in N$ has a unique successor $n' \in N$; we're used to calling the successor $n + 1$. No natural number is the successor of two different natural numbers. The number we call 1 is the only element of N that is not a successor of any other element of **N**. All the other elements of **N** can be produced by forming successors: 2 is the successor of 1, 3 is the successor of 2, 4 is the successor of 3, and so on.

There's a wealth of information in the preceding paragraph. It includes the basis for an important logical principle called **mathematical induction.** In its simplest form, it says that if a set of natural numbers contains 1 and also contains the successor of each of its elements, then the set is all of **N**. That allows us to define operations on **N** one element at a time and provides us with a powerful method for establishing their properties. For example, consider the addition of natural numbers. What is meant by $m + n$? Given $m \in \mathbf{N}$, we can define the sum $m + 1$ to be the successor of m, $m + 2$ to be the successor of $m + 1$, and so on. Once we've defined $m + n$, we can define m plus the successor of n to be the successor of $m + n$. So the set of all n for which this process defines $m + n$ is a set that contains 1 and the successor of each of its elements. According to the principle of mathematical induction, this process defines $m + n$ for all natural numbers n.

If we're ambitious, we can use this formal definition of the addition of natural numbers to develop rigorous proofs of the familiar laws for addition. This was all worked out by an Italian mathematician, Giuseppe Peano, in the late nineteenth century. We won't pursue his development further, since our understanding of the laws of arithmetic is already adequate for our purposes. But we will return to the principle of mathematical induction repeatedly. It appears in two forms: in inductive definitions, as above, and in inductive proofs. Sometimes it appears in the form of another principle we use for working with sets of natural numbers: each nonempty subset of **N** contains a smallest element. Mathematicians refer to this principle by saying that **N** is a **well-ordered set.**

To improve our understanding of mathematical induction, let's use it to prove that **N** is well-ordered. We need to show that every subset of **N** is either empty or has a smallest element. We'll do so by assuming only that $E \subset \mathbf{N}$ and that E has no smallest element, then proving that E must be the empty set. Our strategy is prove that $n \in \mathbf{N}$ implies that $n \notin E$, so that E can't have any elements. We indicate this by writing $E = \emptyset$, the standard symbol for the empty set. It's easy to see why $1 \notin E$; whenever 1 is in a set of natural numbers, it is necessarily the smallest number in the set. After we've learned that $1 \notin E$, we can deduce that $2 \notin E$ because 2 is the smallest number in any set of natural numbers that contains 2 but not 1. This line of reasoning can be continued, and that's how induction comes in. Call I_n the set $\{1, 2, \ldots, n\}$ with $n \in \mathbf{N}$ so that $I_n \cap E$ (the intersection of I_n and E) represents all the numbers in E that are included in the first n natural numbers. Whenever n has the property that $I_n \cap E = \emptyset$, we can use the assumption that E has no smallest element to deduce that $n + 1$ can't be in E, and therefore $I_{n+1} \cap E = \emptyset$ as well. Since we do know that $I_1 \cap E = \emptyset$, the inductive principle guarantees that $I_n \cap E = \emptyset$ for all $n \in \mathbf{N}$.

The next development is to use addition to define the inverse operation, subtraction, with $m - n$ initially defined only for m larger than n. To make subtraction of arbitrary natural numbers possible, we enlarge the system $\mathbf{N}$ to form $\mathbf{Z}$, the system of all integers (the symbol $\mathbf{Z}$ comes from the German word *Zahl* for numeral). The system $\mathbf{Z}$ includes $\mathbf{N}$, 0, and each negative integer $-n$ with $n \in \mathbf{N}$. As in $\mathbf{N}$, each integer $j \in \mathbf{Z}$ has a unique successor in $\mathbf{Z}$, but the difference is that each has a unique predecessor as well. If we try to run through the elements of $\mathbf{Z}$ one at a time, we see that we can't list them in increasing order; there's no place we can start without leaving out lower integers. But if we don't try to put them in numerical order, there are lots of ways to indicate them in a list, such as

$$\mathbf{Z} = \{0, 1, -1, 2, -2, \ldots, n, -n, \ldots\}.$$

As we do mathematics, we often consider the elements of a set one at a time, with both the starting point and the progression clearly defined. When we do so, we are working with a mathematical object called a **sequence**, the mathematical term for a numbered list. On an abstract level, a sequence in a set S is a special sort of indexed collection $\{x_n : n \in I\}$ with $x_n \in S$ for each $n \in I$. The indexed collection is a sequence if the index set I is a subset of $\mathbf{Z}$ with two special properties: it has a least element, and for each element n of I either $n + 1 \in I$ or n is the greatest element of I. That lets us use mathematical induction to define sequences or to prove things about specific ones, as in our proof that $\mathbf{N}$ is a well-ordered set. In the abstract, we prefer index sets with 1 the least element, so that x_1 is the first element of the sequence and x_n is the nth element. Logically, we could just as well make this preference a requirement. But in practice it may be more natural to do something else, and that's the reason we defined sequences the way we did. For example, we might define a sequence by calling a_n the coefficient of x^n in the expansion of $\left(x^{-1} + 1 + x\right)^7$, making a_{-7} the first term and a_7 the last.

Since the index set of a sequence is completely specified by giving its least element and either giving its largest element or saying that it has none, sequences are often indicated by giving this information about the index set in place of an explicit definition of it. For example, $\{x_n\}_{n=0}^{7}$ indicates a sequence with index set $\{0, 1, 2, 3, 4, 5, 6, 7\}$, and $\{x_n\}_{n=1}^{\infty}$ indicates a sequence with index set all of $\mathbf{N}$.

The rational number system $\mathbf{Q}$ contains all fractions m/n with $n \neq 0$ and $m, n \in \mathbf{Z}$. The symbol $\mathbf{Q}$ is used because rational numbers are quotients of integers. Different pairs of integers can be used to indicate the same rational number. The rule is that

$$\frac{m}{n} = \frac{m^*}{n^*} \quad \text{if } mn^* = m^*n.$$

It's at this stage that the difference between mathematical operations and other things we might do with mathematical symbols becomes important. When we perform a mathematical operation on a rational number x, the result isn't supposed to depend on the particular pair of integers we use to represent x, even though we may use them to express the result. For example, we may add 2 to the rational number $\frac{m}{n}$ and express the result as $\frac{2n+m}{n}$; this is a legitimate mathematical operation. On the other hand, converting $\frac{m}{n}$ to $\frac{2+m}{n}$ doesn't correspond to any process we can call a mathematical operation on the rational number $\frac{m}{n}$, it's just something else we can do with m and n.

The rational numbers can also be identified with their decimal expansions. When the quotient of two integers is computed using the standard long division algorithm, the resulting decimal either terminates or repeats. Conversely, any terminating decimal or repeating decimal represents a rational number; that is, it can be written as the quotient of two integers. Obviously, a terminating decimal can be written as an integer divided by a power of 10. There's also a way to write a repeating decimal as a terminating decimal divided by a natural number of the form $10^n - 1$. There's a clever trick involved: if the repeating part of x has n digits, then $10^n x$ has the same repeating part, and so the repeating parts cancel out when we compute $(10^n - 1)x$ as $10^n x - x$.

When considered in numerical order, elements of $\mathbf{Q}$ have neither immediate successors nor predecessors. No matter which two distinct rational numbers we specify, there are always additional rational numbers between them. So we never think of one rational number as being next to another. If we disregard the numerical order of the rationals, it is still possible to come up with a sequence that includes every element in $\mathbf{Q}$, but we have no reason to pursue that here.

It's common to think of the rational numbers as arranged along the x-axis in the coordinate plane. The rational number $\frac{m}{n}$ corresponds to the intersection of the x-axis with the line through the lattice points $(m, n - 1)$ and $(0, -1)$. Although it looks like one can fill the entire x-axis with such points, it's been known since the time of Pythagoras that some points on the line don't correspond to any rational number. For example, if we form a circle by putting its center at $(0,0)$ and choosing the radius to make the circle pass through $(1,1)$, then the circle crosses the x-axis twice but not at any point in $\mathbf{Q}$. Extending $\mathbf{Q}$ to a larger set of numbers that corresponds exactly to all the points on the line produces the real number system $\mathbf{R}$. The fact that there are no missing points is the geometric version of the completeness property of the real numbers, a property we'll see a great deal more of.

Calculus deals with variables that take their values in $\mathbf{R}$, so a reasonably good understanding of $\mathbf{R}$ is needed before one can even comprehend the possibilities. The real number system is far more complicated than one might expect. Our computational algorithms can't really deal with complete decimal representations of real numbers. For example, decimal addition is supposed to begin at the rightmost digit. Mathematicians have a simple way to get around this difficulty; we just ignore it. We simply indicate arithmetic operations in $\mathbf{R}$ with algebraic notation, treating the symbols that represent real numbers in much the same way we treat unknowns.

EXERCISES

1. Use the principle of mathematical induction to define 2^n for all $n \in \mathbf{N}$. (A definition using this principle is usually said to be given **inductively** or **recursively**.)

2. The sequence $\{s_n\}_{n=1}^{\infty}$ whose nth term is the sum of the squares of the first n natural numbers can be defined using Σ-notation as

$$s_n = \sum_{k=1}^{n} k^2.$$

It can also be defined recursively by specifying

$$s_1 = 1 \quad \text{and} \quad s_{n+1} = s_n + (n+1)^2 \quad \text{for all } n \in \mathbf{N}.$$

Use the inductive principle to prove that $s_n = \frac{1}{6}n(n+1)(2n+1)$ for all $n \in \mathbf{N}$.

3. Why is it impossible to find a sequence that includes every element of $\mathbf{Z}$ with all the negative integers preceding all the positive ones? Suggestion: given a sequence in $\mathbf{Z}$ that includes 1 as a term, explain why the preceding terms can't include all the negative integers.

4. Suppose that m and n are nonnegative integers such that $m^2 = 2n^2$. Use simple facts about odd and even integers to show that m and n

are both even and that $m/2$ and $n/2$ are also nonnegative integers with $(m/2)^2 = 2(n/2)^2$. Why does this imply that $m = n = 0$?

2   INTERVALS

Many of the features of $\mathbf{N}$, $\mathbf{Z}$, $\mathbf{Q}$, and $\mathbf{R}$ are described in terms of the numerical order of their elements. Consequently, we often work with sets of numbers defined in terms of numerical order; the simplest such sets are the intervals. A nonempty set of numbers is called an **interval** if it has the property that every number lying between elements of the set must also be an element. We're primarily interested in working with sets of real numbers, so when we say a nonempty set $I \subset \mathbf{R}$ is an interval, it means that we can prove a given $x \in \mathbf{R}$ is an element of I by simply finding $a, b \in I$ with $a < x < b$. However, I may well contain numbers less than a or greater than b, so to show that a second given real number x' is also in I we might well need to find a different pair of numbers $a', b' \in I$ with $a' < x' < b'$.

We often identify intervals in terms of their **endpoints**. The sets

$$(c, d) = \{x \in \mathbf{R} : c < x < d\} \quad \text{and} \quad [c, d] = \{x \in \mathbf{R} : c \leq x \leq d\}$$

are familiar examples. The custom in the United States is to use a square bracket to indicate that the endpoint is included in the interval and a parenthesis to indicate that it isn't. We should remember that intervals can also be specified in many ways that do not involve identifying their endpoints; our definition doesn't even require that intervals have endpoints.

Instead of simply agreeing that (c, d) and $[c, d]$ are intervals because that's what we've always called them, we should see that they really do satisfy the definition. That's the way mathematics is done. In the case of (c, d), we should assume only that $a, b \in (c, d)$ and $a < x < b$, and then find a reason why x must also be in (c, d). The transitive law for inequalities provides all the justification we need: for $a, b \in (c, d)$ we must have $c < a$ and $b < d$, and then $a < x < b$ implies that $c < x < d$. Similar considerations explain why $[c, d]$ is an interval.

Somewhat surprisingly, a set consisting of a single real number is an interval. When I has a single element it is nonempty. Since we can't possibly find numbers $a, b \in I$ with $a < b$ we need not worry whether every x between elements of I satisfies $x \in I$. Such an interval, called a **degenerate interval**, is by no means typical. Note that $[c, c]$ always represents a degenerate interval, but (c, c) does not represent an interval since it has no elements.

We say that an interval is **closed on the right** if it contains a greatest element and **open on the right** if it doesn't. We also say that an interval is **closed on the left** if it contains a least element and **open on the left** if it doesn't. A **closed interval** is an interval that is both closed on the right and closed on the left. It's easy to see that every closed interval must have the form $[c, d]$ with c its least element and d its greatest. An **open interval** is an interval that is both open on the right and open on the left. While every interval of the form (c, d) is an open interval, there are other possibilities to consider. We'll return to this point later in this section.

We say that an interval I is **finite** (or **bounded**) if there is a number M such that every $x \in I$ satisfies $|x| \le M$. Every closed interval is finite because every $x \in [c, d]$ satisfies $|x| \le |c| + |d|$. Open intervals may be either finite or infinite. The set P of all positive real numbers is an example of an infinite open interval. With infinite intervals it's convenient to use the symbol $-\infty$ or ∞ in place of an endpoint; for example, we write $(0, \infty)$ for P. We don't use a square bracket next to $-\infty$ or ∞ because these symbols do not represent elements of any set of real numbers.

We often use intervals to describe the location of a real number that we only know approximately. The shorter the interval, the more accurate the specification. While it can be very difficult to determine whether two real numbers are exactly equal or just very close together, mathematicians generally assume that such decisions can be made correctly; that's one of the basic principles underlying calculus. In fact, we assume that it's always possible to find an open interval that separates two given unequal numbers. We'll take that as an axiom about the real numbers, rather than search for some other principle that implies it.

AXIOM 1: *Given any two real numbers a and b, either $a = b$ or there is an $\varepsilon > 0$ such that $a \notin (b - \varepsilon, b + \varepsilon)$.*

Of course, any ε between 0 and $|b - a|$ should work when $a \ne b$; one of the things the axiom says is that $|b - a|$ is positive when $a \ne b$. In particular, two different real numbers can never be thought of as being arbitrarily close to each other. That's why we say, for example, that the repeating decimal $0.\overline{9}$ and the integer 1 are equal; there is no positive distance between them. We often use the axiom to prove that two real numbers a and b are equal by proving that $a \in (b - \varepsilon, b + \varepsilon)$ for every positive ε.

While a single interval may represent an inexact specification of a real number, we often use sequences of intervals to specify real numbers exactly. For example, it is convenient to think of a nonterminating decimal as specifying a real number this way. When we read through the first n digits to the right of the decimal point and ignore the remaining ones, we're specifying a closed interval I_n of length 10^{-n}. For example, saying that the decimal expansion for π begins with 3.1415 is equivalent to saying that π is in the interval $[3.1415, 3.1416]$. The endpoints of this interval correspond to the possibilities that the ignored digits are all 0 or all 9. The complete decimal expansion of π would effectively specify an infinite sequence of intervals, with π the one real number in their **intersection**, the mathematical name for the set of numbers common to all of them. With $\{I_n\}_{n=1}^{\infty}$ being the sequence of closed intervals corresponding to the complete decimal expansion of π, we write

$$\bigcap_{n=1}^{\infty} I_n = \{\pi\}\,;$$

that is, the intersection of all the intervals I_n in the sequence is the set whose only element is π.

In using the intersection of a sequence of intervals to define a real number with some special property, there are two things we have to check. The intersection can't be empty, and every other real number except the one we're defining must be excluded. There are some subtleties in checking these conditions, so to simplify the situation we usually try to work with sequences $\{I_n\}_{n=1}^{\infty}$ such that each interval I_n in the sequence includes the next interval I_{n+1} as a subinterval. We call such a sequence a **nested sequence** of intervals; the key property is that $I_{n+1} \subset I_n$ for all $n \in \mathbf{N}$. For any nested sequence of intervals,

$$\bigcap_{n=1}^{m} I_n = I_m \quad \text{for all } m \in \mathbf{N},$$

so we can at least be sure that every finite subcollection of the intervals in the sequence will have a nonempty intersection. However, $\bigcap_{n=1}^{\infty} I_n$ can easily be empty, even for nested intervals; defining $I_n = (n, \infty)$ provides an easy example. We can rule out such simple examples if we restrict our attention to closed intervals. By using any of the common statements of the completeness property it is possible to show that the intersection of a nested sequence of closed intervals can't be the empty set. But instead of proving this as a theorem, we'll take it as an axiom; it is easier to understand than the assumptions we would need to make to prove it. It's also fairly easy to work with.

AXIOM 2: *Every nested sequence of closed intervals in* **R** *has a nonempty intersection.*

Now let's look at the second part of the problem of using a sequence of intervals to define a real number. How can we know that the sequence excludes every other real number except the one we're trying to define? Thanks to Axiom 1, that's easy. Let's say $\{[a_n, b_n]\}_{n=1}^{\infty}$ is a nested sequence of closed intervals, with r a point in $\bigcap_{n=1}^{\infty}[a_n, b_n]$. Given any $x \neq r$, we know there is an $\varepsilon > 0$ such that $x \notin (r - \varepsilon, r + \varepsilon)$, and then x can't be in any interval that includes r but has length less than ε. So our nested sequence $\{[a_n, b_n]\}_{n=1}^{\infty}$ can be used to define a real number if the sequence has the property that for each $\varepsilon > 0$ there is an m with $b_m - a_m < \varepsilon$. We'll indicate this condition by writing $b_n - a_n \to 0$, anticipating the notation of Chapter 3 for the limit of a sequence. Note that when we write $b_n - a_n \to 0$, we are indicating a property of the sequence as a whole, not a property of a particular term in the sequence. It's also helpful to recognize that for a nested sequence of closed intervals,

$$b_n - a_n \leq b_m - a_m \quad \text{for all } n \geq m.$$

Earlier we explained how a nonterminating decimal expansion could be interpreted as a sequence of intervals. Note that the intervals form a nested sequence of closed intervals. The length of the nth interval in the sequence is 10^{-n}, and obviously $10^{-n} \to 0$. After all, we can use the decimal expansion of any given $\varepsilon > 0$ to find n with $10^{-n} < \varepsilon$. So we see that our axioms guarantee that every nonterminating decimal expansion defines exactly one real number.

Here is a more surprising consequence of Axiom 2: it is impossible to write a sequence of real numbers such that every real number in a closed nondegenerate interval appears as a term. This fact about the real number system was discovered by the German mathematician Georg Cantor in the nineteenth century. He assumed that a sequence of real numbers had been given, and then used the given sequence to produce a real number that could not have been included. This is easy to do inductively. First choose a closed nondegenerate interval that excludes the first number in the sequence, then successively choose closed nondegenerate subintervals that exclude the successive numbers in the sequence. Then any number in all the intervals can't be in the sequence.

We often locate real numbers by the **bisection method**, using Axiom 2 as the theoretical basis. We start with a closed interval and successively take either the closed right half or the closed left half of the interval just selected. That is, if $I_n = [a_n, b_n]$, then the next interval I_{n+1} is $[a_n, m_n]$ or

$[m_n, b_n]$, where m_n is the midpoint of I_n. This always makes $b_n - a_n \to 0$, so however we decide which halves to choose we will wind up with exactly one real number common to all the intervals.

We'll conclude this section by using the bisection method to prove a theorem describing all open intervals.

THEOREM 2.1:   *Every finite open interval has the form (a, b) for some pair of real numbers a, b with $a < b$. Every other open interval has one of the forms $(-\infty, \infty)$, (a, ∞), or $(-\infty, b)$.*

☐   *Proof:*   Let I be an open interval. Since our definition of interval requires that $I \neq \emptyset$, we may assume that we have a real number $c \in I$. Since I is open, we know that c is neither the least nor the greatest number in I. We'll examine the portions of I to the right and left of c separately, beginning with the right.

Either $(c, \infty) \subset I$ or there is a real number $d > c$ with $d \notin I$. In the latter case, we prove that there is a real number $b > c$ with both $(c, b) \subset I$ and $I \cap [b, \infty) = \emptyset$, making b the right endpoint for I. To produce b, we let $I_1 = [c, d]$, and define a nested sequence of closed intervals by the bisection method. Our rule is simple: when the midpoint of I_n is in the original interval I, we define I_{n+1} to be the closed right half of I_n. When the midpoint isn't in I, we define I_{n+1} to be the closed left half of I_n. Then each interval in our nested sequence has its left endpoint in I and its right endpoint outside I. It's important to note that since I is an interval we know that every point between c and any one of the left endpoints is in I, while no point to the right of any one of the right endpoints can be in I.

Now we define b to be the one real number common to all the closed intervals in the nested sequence just constructed. Since the first interval was $[c, d]$, we know that $b \geq c$. Note that if $x > b$ then x can't be in I because when the length of I_n is less than $x - b$ we know x must be to the right of the right endpoint of I_n. Since I has no greatest element, we can also conclude that $b \notin I$ and so $I \cap [b, \infty) = \emptyset$. Since $c \in I$, we must have $c < b$. Our next task is to show $(c, b) \subset I$. Given any x in (c, b), we know x can't be in I_n when the length of I_n is less than $b - x$, so x must be to the left of the left endpoint of I_n. That puts x between c and the left endpoint of I_n, so $x \in I$ and we've proved that $(c, b) \subset I$.

Now we consider numbers below c. Either $(-\infty, c) \subset I$ or there is a number $e < c$ with $e \notin I$. In the latter case, we can argue as above to produce a real number $a < c$ with $(a, c) \subset I$ but $I \cap (-\infty, a] = \emptyset$.

When we sort through the various possibilities, we conclude that every open interval must have one of the four forms given in the statement of the theorem. For example, if $(c, \infty) \subset I$ and also $(a, c) \subset I$ but

$I \cap (-\infty, a] = \emptyset$, then we can conclude that $I = (a, \infty)$. We finish the proof by noting that (a, b) is the only one of these four forms that represents a finite interval.   ∎

Theorem 2.1 is actually a deep result about **R**; it's another version of the completeness principle. See Exercise 9 below for a brief development of this idea. One of the recurring problems we'll encounter is to produce a real number with some special property, and one of the ways to do so is to identify it as an endpoint of a carefully defined open interval. However, in the next section we'll develop procedures that are somewhat easier to use.

EXERCISES

5. Use the definition of interval to prove that if I and J are intervals with $I \cap J \neq \emptyset$, then both $I \cap J$ and $I \cup J$ are intervals.

6. For $I_n = \left[\frac{1}{n}, \frac{2}{n}\right]$, show that $\bigcap_{n=1}^{\infty} I_n = \emptyset$. What hypothesis in the nested interval axiom does this sequence fail to satisfy?

7. For $I_n = \left(0, \frac{1}{n}\right)$, show that $\bigcap_{n=1}^{\infty} I_n = \emptyset$. What hypothesis in the nested interval axiom does this sequence fail to satisfy?

8. Let $\{d_n\}_{n=1}^{\infty}$ be the sequence in $\{0, 1\}$ defined by $d_n = 1$ when n is a perfect square and $d_n = 0$ when n is not a perfect square. Prove that $0.d_1 d_2 d_3 \ldots d_n \ldots$ is a nonterminating, nonrepeating decimal.

9. For $\{[a_n, b_n]\}_{n=1}^{\infty}$ an arbitrary nested sequence of closed intervals, show that $\bigcup_{n=1}^{\infty} (-\infty, a_n)$ is an open interval, and prove that its right endpoint must be an element of $\bigcap_{n=1}^{\infty} [a_n, b_n]$.

10. Use the bisection method to prove that 2 has a positive square root in **R**. The crucial fact is that if $0 < x < y$, then $x^2 < y^2$. Begin your search for $\sqrt{2}$ in the interval $[1, 2]$.

11. Without knowing that $\sqrt{2}$ exists, how could you define an open interval whose left endpoint is $\sqrt{2}$?

3 SUPREMA AND INFIMA

Working with intervals that fail to have a least or greatest element is usually easy, since an endpoint of the interval can be an adequate substitute. For sets of real numbers that are not intervals, it's often useful to have numbers that play the role of endpoints. In place of the left endpoint of E, we might use the greatest a such that $E \subset [a, \infty)$, and in place of the right endpoint, we might use the least b such that $E \subset (-\infty, b]$. As the case of open intervals illustrates, such numbers aren't necessarily elements of the given set, so we can't quite call them least or greatest elements, but that's almost the concept we have in mind. To suggest that notion without quite saying it, we make use of the Latin words *infimum* and *supremum*, meaning lowest and highest, and indicate them symbolically as inf E and sup E. The familiar English words inferior and superior come from the same Latin roots, so the meanings of these terms are easy to keep straight. Here's a formal definition of these terms as we will use them.

DEFINITION 3.1: For E a nonempty set of real numbers, we say that the real number b is the **supremum** of E, and write $b = \sup E$, if every $x \in E$ satisfies $x \leq b$ but for each $c < b$ there is an $x \in E$ with $x > c$. Similarly, we say that the real number a is the **infimum** of E, and write $a = \inf E$, if every $x \in E$ satisfies $x \geq a$ but for each $c > a$ there is an $x \in E$ with $x < c$.

This definition deserves to be examined quite carefully. Saying that every $x \in E$ satisfies $x \leq b$ is equivalent to the set-theoretic inclusion $E \subset (-\infty, b]$. We describe this relationship by saying that b is an upper bound for E. Saying that some $x \in E$ satisfies $x > c$ is then equivalent to saying that c is not an upper bound for E. Thus, according to the definition, the supremum of a set is an upper bound with the property that no smaller number is an upper bound for the set, so the supremum of a set is sometimes called the **least upper bound**. Similarly, the infimum is sometimes called the **greatest lower bound**. But it's all too easy to confuse least upper bounds with least elements or greatest lower bounds with greatest elements, so we'll stick with the names supremum and infimum.

By the way, as we compare values of real numbers and describe the results, we often use the words *less, least, greater,* or *greatest* rather than *smaller, smallest, larger,* or *largest.* When we're comparing positive quantities it makes little difference, and the latter terms are more common in everyday use. But while we easily recognize that $-1,000,000$ is less than -1, most people don't think of it as smaller; the word *smaller* suggests being closer to zero instead of being farther to the left. In general, we should avoid using any terms that suggest the wrong concept even if we know our use of the terms is correct; being right but misunderstood isn't much better than being wrong.

Returning to the notion of supremum and infimum, there's an obvious question: why should a set of real numbers have a supremum or infimum? In fact, some hypotheses about the nature of the set are essential, but according to the theorem below, they are remarkably simple.

THEOREM 3.1: *Every nonempty set of real numbers that has an upper bound has a supremum, and every nonempty set of real numbers that has a lower bound has an infimum.*

Some mathematicians prefer to call this theorem an axiom because it's another version of the completeness property for $\mathbf{R}$. It can serve as a logical substitute for our axiom about nested sequences of closed intervals; when this theorem is assumed to be true, our axiom can be derived as a consequence. Conversely, we can and do use our axiom to prove the theorem. We use bisection to locate the supremum and infimum with almost the same argument we used to prove our theorem about the form of open intervals.

$\square$ *Proof:* Let's begin by assuming that E is a nonempty set of numbers, and that E has an upper bound b_1. There must be a number $x \in E$, and if we choose a number $a_1 < x$ then a_1 is not an upper bound for E. Thus $I_1 = [a_1, b_1]$ is a closed interval whose right endpoint is an upper bound for E, but whose left endpoint is not. Using bisection and mathematical induction, we can produce a nested sequence of closed intervals $\{I_n\}_{n=1}^{\infty}$, each interval having these same properties. For $I_n = [a_n, b_n]$, its midpoint is $m_n = \frac{1}{2}(a_n + b_n)$. We define

$$I_{n+1} = \begin{cases} [a_n, m_n], & m_n \text{ an upper bound for } E \\ [m_n, b_n], & m_n \text{ not an upper bound for } E. \end{cases}$$

Then each I_n is a closed interval containing the next interval in the sequence and the lengths satisfy $b_n - a_n \to 0$. Once again, there is exactly one real number common to all these intervals; let's call it s.

Now let's see why $s = \sup E$. For any given $x > s$, there is an interval I_n in the sequence with length less than $x - s$, so its right endpoint is to the left of x. That shows that x is greater than an upper bound for E, so $x \notin E$ when $x > s$. Hence every $x \in E$ satisfies $x \leq s$. On the other hand, given any $y < s$, there is an interval I_n with its left endpoint to the right of y. Since that endpoint isn't an upper bound for E, we see that y isn't either; some $z \in E$ satisfies $y < z$. We've proved that $s = \sup E$.

The same sort of argument is used to prove that every nonempty set of real numbers with a lower bound has an infimum. We leave that proof to the reader. $\blacksquare$

Our first use of the existence of suprema and infima will be to establish a useful substitute for Axiom 2. There are times when it's convenient to define a real number as the one point common to all the intervals in some collection when the collection is not a nested sequence. The theorem below is a real help in such situations. In particular, we'll use it in Chapter 5 to help develop the theory of integration.

THEOREM 3.2: *Let $\mathcal{I}$ be a nonempty family of closed intervals such that every pair of intervals in $\mathcal{I}$ has a nonempty intersection. Then there is at least one point that is common to every interval in $\mathcal{I}$.*

$\square$ *Proof:* For L the set of left endpoints of intervals in $\mathcal{I}$, we'll prove that L has a supremum and that $\sup L$ is in every interval in $\mathcal{I}$. We're given that $\mathcal{I}$ is a nonempty family of closed intervals; let's consider an arbitrary $[a, b] \in \mathcal{I}$. Clearly $a \in L$, and if we can prove that b is an upper bound for L then we'll know that L has a supremum. We'll also know that

$$a \leq \sup L \leq b.$$

Given any other interval $[c, d] \in \mathcal{I}$, its intersection with $[a, b]$ is nonempty. So there must be an $x \in \mathbf{R}$ with

$$x \in [a, b] \cap [c, d].$$

That means both

$$a \leq x \leq b \quad \text{and} \quad c \leq x \leq d,$$

so

$$c \leq x \leq b.$$

Thus $c \leq b$ for every $[c, d] \in \mathcal{I}$, so b is an upper bound for L and therefore $\sup L \in [a, b]$. Since $[a, b]$ represents an arbitrary interval in $\mathcal{I}$, this proves that $\sup L$ is in every interval in $\mathcal{I}$. ∎

Another useful notion closely related to the supremum and infimum of a set is the set's **diameter**. Geometers use the word *diameter* to refer to a line segment that joins two points on a circle and passes through the center of the circle. Of all the line segments joining two points on a given circle, the diameters have the greatest length; that length is what most of us mean by the diameter of a circle. Sets of real numbers don't look very much like circles, but we can still talk about the length of a line segment connecting two numbers in the set. Of course, we calculate that length by subtracting the smaller number from the larger, so the set of all lengths of line segments linking two points in the set E of real numbers is

$$L = \{|x - y| : x, y \in E\}.$$

If L has a supremum, we call that supremum the diameter of E. It's not too hard to see that the diameter of E can be defined if and only if E is a nonempty set that has both an upper and a lower bound. The diameter of a closed interval is simply its length, and this is one of the cases where the set of lengths has a largest element. The theorem below extends this formula for the diameter to more general sets.

THEOREM 3.3: *Let E be a nonempty subset of* **R**. *If E has both an upper and a lower bound, then*

$$\operatorname{diam} E = \sup E - \inf E.$$

□ *Proof:* As with most formulas involving suprema and infima, several steps are required to verify it. We need to show that $\sup E - \inf E$ satisfies the definition of $\sup L$. First, we note that $E \subset [\inf E, \sup E]$, so for every possible choice of $x, y \in E$ we have

$$|x - y| \le \sup E - \inf E,$$

proving that $\sup E - \inf E$ is an upper bound for L. That's the first step. To complete the proof, we show that for every $b < \sup E - \inf E$, there is an element of L that is greater than b.

Given $b < \sup E - \inf E$, we call

$$\varepsilon = \sup E - \inf E - b.$$

Then $\varepsilon > 0$, so

$$\sup E - \frac{\varepsilon}{2} < \sup E \quad \text{and} \quad \inf E < \inf E + \frac{\varepsilon}{2}.$$

Then by the definition of $\sup E$ and $\inf E$, there are $x, y \in E$ such that

$$x > \sup E - \frac{\varepsilon}{2} \quad \text{and} \quad y < \inf E + \frac{\varepsilon}{2}.$$

These elements of E satisfy

$$|x - y| \ge x - y > \sup E - \inf E - \varepsilon,$$

and the quantity on the right is just b. ∎

In many problems we find that we need the supremum or infimum of a set of sums of numbers, with each term in the sum representing an element of some simpler set. As the theorem below indicates, we can deal with such problems by analyzing the component sets one at a time.

THEOREM 3.4: *Suppose that* $\{E_1, E_2, \ldots, E_n\}$ *is a collection of sets of numbers and that each of the sets has a supremum. Then*

$$\sup\left\{\sum_{k=1}^{n} x_k : x_1 \in E_1, x_2 \in E_2, \ldots, x_n \in E_n\right\} = \sum_{k=1}^{n} \sup E_k.$$

Similarly, if each of the sets E_k has an infimum, then the infimum of the set of sums is the sum of the infima.

☐ *Proof:* Once we understand the case $n = 2$, it's a simple matter to complete the proof by using induction. For the inductive step, we note that when our collection of sets is expanded from $\{E_1, \ldots, E_n\}$ to $\{E_1, \ldots, E_n, E_{n+1}\}$, we can regroup $\sum_{k=1}^{n+1} x_k$ as $(\sum_{k=1}^{n} x_k) + x_{n+1}$, with each of these two numbers representing an element of a set with a known supremum. With that in mind, let's simply assume that E and F are sets of numbers and that both sets have suprema. Then we show that

$$\sup\{x + y : x \in E \text{ and } y \in F\} = \sup E + \sup F.$$

By assumption, neither E nor F can be the empty set, so

$$S = \{x + y : x \in E \text{ and } y \in F\}$$

is not the empty set. Every $x \in E$ satisfies $x \leq \sup E$ and every $y \in F$ satisfies $y \leq \sup F$, so every $x + y \in S$ must satisfy

$$x + y \leq \sup E + \sup F.$$

On the other hand, given any $c < \sup E + \sup F$, if we call

$$\varepsilon = \sup E + \sup F - c,$$

then there must be an $x_c \in E$ with $x_c > \sup E - \varepsilon/2$ and there must be a $y_c \in F$ with $y_c > \sup F - \varepsilon/2$. So we have

$$x_c + y_c \in S \quad \text{and} \quad x_c + y_c > \sup E + \sup F - \varepsilon = c,$$

proving that $\sup E + \sup F$ is the supremum of S. ∎

Mathematicians often introduce sets without having a very good idea how to determine what all the elements are, just as they use letters to represent numbers whose precise values they may not know how to compute. We should think of $\sup E$, $\inf E$, and $\operatorname{diam} E$ in much the same spirit. Determining their exact values requires considering the numbers in a set all at once, not just one at a time, and for complicated sets that may be harder than anything we know how to do. That's good news for mathematicians; the difficulty in finding these values makes the symbols for them more useful. When we use these symbols to formulate and prove relationships, they help us organize complex systems into simpler patterns, neatly avoiding the complexity of the components. That's the sort of thing that makes abstract mathematics a practical skill, not just an intellectual exercise.

EXERCISES

12. Complete the proof that if E is a nonempty set of real numbers and E has a lower bound, then E has an infimum.

13. Provide an argument showing that if E and F are sets of real numbers and each has an infimum, then

$$\inf\left\{x + y : x \in E \text{ and } y \in F\right\} = \inf E + \inf F.$$

14. Theorem 2.1 can also be used to prove the existence of suprema; this exercise shows how. Suppose that $E \subset \mathbf{R}$, and let B be the set of all upper bounds of E. Without assuming that E has a supremum, show that

$$U = \bigcup_{b \in B} (b, \infty)$$

defines an open interval as long as $B \neq \emptyset$. Then show that when $E \neq \emptyset$, the left endpoint of U satisfies the definition of $\sup E$.

15. When $E \subset \mathbf{R}$, it is convenient to define the set

$$-E = \{-x : x \in E\} = \{y : -y \in E\}.$$

Prove that E has a supremum if and only if $-E$ has an infimum and that $\inf(-E) = -\sup E$. **Note:** The second part of a statement like this is generally assumed to include the first part implicitly; the identity of two expressions requires both to be defined in the same cases. Exceptions should be noted explicitly.

16. Find the diameter of the set $\left\{\frac{1}{n} : n \in \mathbf{N}\right\}$.

4 EXACT ARITHMETIC IN R

Exact arithmetic is standard operating procedure in the rational number system $\mathbf{Q}$; the rules are

$$\frac{m}{n} + \frac{m^*}{n^*} = \frac{mn^* + nm^*}{nn^*}, \frac{m}{n} \cdot \frac{m^*}{n^*} = \frac{mm^*}{nn^*}, \text{ and } \frac{m}{n} \bigg/ \frac{m^*}{n^*} = \frac{mn^*}{nm^*}$$

Consequently, the familiar rules of algebra implicitly assume that the indicated arithmetic operations are performed exactly. Of course, when we do arithmetic using decimals, we often wind up with approximate values instead of exact values. If we're not careful that can cause some problems; even the best computers are subject to errors introduced when decimal values are rounded off. Here's a simple example of the sort commonly used to introduce calculus operations.

EXAMPLE. Compute representative values of $\frac{(1+h)^2 - 1}{h}$ for h in a small interval $(0, \varepsilon)$.

In terms of exact arithmetic, the fraction is exactly equal to $2 + h$, so the resulting values are all close to 2. But if we ask a computer to evaluate the original expression with $h = 10^{-50}$, the computer is likely to return a value of 0. While most computers are capable of recognizing $10^{-50} \neq 0$, they find it difficult to tell that $\left(1 + 10^{-50}\right)^2 \neq 1$ because in processing numerical values digitally they usually ignore digits too far to the right of the first nonzero one. Consequently, the computer will treat the original fraction as 0 divided by a nonzero number.

For arithmetic operations in the real number system $\mathbf{R}$, the usual rules of algebra are valid, assuming that we do exact arithmetic. But if a number is known to us only as a description of a process for forming a nested sequence of closed intervals, how are we supposed to do arithmetic with it? The way we answer such a question depends on what we understand a real number to be. The fact is that real numbers may only be understood as approximations, and given any approximation we can imagine a context in which that approximation won't be adequate.

Mathematicians are professional pessimists. We tend to believe that anything that can possibly go wrong certainly will go wrong sooner or later, probably sooner than expected, and with worse consequences than previously imagined. So to specify a real number when the exact value can't be conveniently determined, we don't settle for a single approximation; we want a whole family of approximations that includes at least one for every purpose. Then when we're told how accurate the approximation needs to be, we can select one that meets the requirements. A nested sequence of closed intervals that includes arbitrarily short intervals is a good example since we can regard it as a family of approximations to the one number a common to all the intervals. When we need an approximation x to a that is accurate to within ε, we simply find one of the intervals that is shorter than ε and choose any x from that interval.

The philosophy in performing arithmetic on real numbers is that approximations to the input can be used to produce approximations to the output. Let's examine how it works on the fundamental arithmetic operations of addition, multiplication, and division. Subtraction is so much like addition that there's no need to treat it separately. The idea is to assume that x and y represent approximations to the real numbers a and b, perform the operation under consideration on x and y as well as on a and b, and then express the difference of the results in terms of $x - a$ and $y - b$. We then analyze this expression and figure out how small $|x - a|$ and $|y - b|$ need to be.

Addition is especially simple. We see that

$$|(x + y) - (a + b)| = |(x - a) + (y - b)|$$
$$\leq |x - a| + |y - b|.$$

To obtain $|(x + y) - (a + b)| < \varepsilon$, we require that both $|x - a| < \delta$ and $|y - b| < \delta$ with $\delta = \frac{1}{2}\varepsilon$.

Multiplication is more complicated. Since

$$xy - ab = (x - a)(y - b) + a(y - b) + b(x - a),$$

we see that

$$|xy - ab| \leq |x - a| \, |y - b| + |a| \, |y - b| + |b| \, |x - a|.$$

If we require both $|x - a| < \delta$ and $|y - b| < \delta$, we have

$$|xy - ab| < \delta^2 + |a|\, \delta + |b|\, \delta = \delta(\delta + |a| + |b|),$$

so we choose $\delta > 0$ to make this last product be less than ε. Its second factor is bounded by $1 + |a| + |b|$ when $\delta < 1$, so we can guarantee that $|xy - ab| < \varepsilon$ by requiring $|x - a| < \delta$ and $|y - b| < \delta$ for any $\delta \in (0, 1)$ that satisfies

$$\delta \leq \frac{\varepsilon}{1 + |a| + |b|}.$$

In particular, we can always use

$$\delta = \frac{\varepsilon}{\varepsilon + 1 + |a| + |b|}$$

to satisfy our requirements.

To approximate a/b by x/y, we first must assume that $|b| \neq 0$. Then we restrict our attention to y with $|y - b| < |b|$ to guarantee that $y \neq 0$. We then find that

$$\frac{x}{y} - \frac{a}{b} = \frac{bx - ay}{by} = \frac{b(x - a) - a(y - b)}{by}. \tag{1.1}$$

If we require both $|x - a| < \delta$ and $|y - b| < \delta$, then we have a simple bound for the numerator in (1.1):

$$|b(x - a) - a(y - b)| \leq |b| |x - a| + |a| |y - b| < (|b| + |a|) \delta.$$

If we also require $0 < \delta < |b|$, then the denominator in (1.1) satisfies

$$|by| = |b| |y| > |b| (|b| - \delta).$$

Hence taking absolute values in (1.1) gives the bound

$$\left| \frac{x}{y} - \frac{a}{b} \right| = \frac{|b(x - a) - a(y - b)|}{|by|} < \frac{(|b| + |a|) \delta}{|b| (|b| - \delta)}.$$

Now we choose δ in $(0, |b|)$ to make sure that this last quantity is at most ε. For $|b| - \delta > 0$, the inequality

$$\frac{(|b| + |a|) \delta}{|b| (|b| - \delta)} \leq \varepsilon$$

is satisfied when

$$(|b| + |a|) \delta \leq |b| (|b| - \delta) \varepsilon = b^2 \varepsilon - |b| \delta \varepsilon,$$

or, equivalently, when

$$(|a| + |b| + |b| \varepsilon) \delta \leq b^2 \varepsilon.$$

Consequently, as long as $b \neq 0$ we can choose

$$\delta = \frac{b^2 \varepsilon}{|a| + |b| + |b| \varepsilon};$$

this also gives us that $0 < \delta < |b|$. Then requiring $|x - a| < \delta$ and $|y - b| < \delta$ will guarantee that $|x/y - a/b| < \varepsilon$, for whatever $\varepsilon > 0$ is given.

It's important to note the form of our answers. For any given $\varepsilon > 0$, we guarantee that the approximate answer is within ε of the exact answer as long the approximate input is within δ of the exact input. The quantitative relationship between δ and ε changes as the arithmetic operation changes, but the logical relationship stays the same. That, in fact, is why we prefer to use the name δ and not the formula we find for it. With just a little imagination, we can then recognize that the same sort of conditions can be developed for more complicated arithmetic operations. The quantitative relationship between δ and ε can be built up in stages, just as the operation is built in stages using the basic building blocks of addition, multiplication, and division.

Here's a simple example showing how it's done. Let's say x, y, and z are approximations to a, b, and c, respectively, and we're going to use $x(y + z)$ as an approximation to $a(b + c)$. We assume that $\varepsilon > 0$ has been given. We need a $\delta > 0$ such that whenever $|x - a|, |y - b|$, and $|z - c|$ are all smaller than δ, we can be certain that $x(y + z)$ is within ε of $a(b + c)$. The last thing we do in calculating $x(y + z)$ is to multiply x and $y + z$, and we know enough about approximate multiplication to know there is a $\delta_1 > 0$ such that

$$|x(y + z) - a(b + c)| < \varepsilon$$

as long as

$$|x - a| < \delta_1 \quad \text{and} \quad |(y + z) - (a + b)| < \delta_1.$$

Now we use δ_1 as another value for ε; our knowledge of approximate addition tells us there is a $\delta_2 > 0$ such that

$$|(y + z) - (a + b)| < \delta_1$$

as long as

$$|y - b| < \delta_2 \quad \text{and} \quad |z - c| < \delta_2.$$

We can therefore choose $\delta > 0$ to be the smaller of δ_1 and δ_2, which is δ_2 in this case.

That's how we interpret exact arithmetic when only approximations are possible. We can guarantee that the output possesses whatever level of accuracy is desired by simply making sure that the input has sufficient accuracy; using exact input is not crucial.

EXERCISES

17. Given a, b with $a + b > 0$, explain why there must a $\delta > 0$ such that $u + v > 0$ for all u, v with $|u - a| < \delta$ and $|v - b| < \delta$.

18. Given a, b, c, d with $cd \neq 0$, explain why whenever $\varepsilon > 0$ is given, there must be a $\delta > 0$ with the property that every x, y, z, and w with $|x - a|, |y - b|, |z - c|$, and $|w - d|$ all less than δ must satisfy

$$\left| \frac{xy}{zw} - \frac{ab}{cd} \right| < \varepsilon.$$

19. For x an approximation to a and $a \neq 0$, the quantity $(x - a)/a$ is called the **relative error**. When x and y are approximations to a and b and xy is used to approximate ab, show how to express the relative error of the product in terms of the relative errors of the factors.

20. When $a \neq 0$ is approximated by x, we use $1/x$ to approximate $1/a$. Express the relative error in the latter approximation in terms of the relative error in the first.

21. For $x \in [a, b]$ and $y \in [c, d]$, it's always true that $x + y \in [a + c, b + d]$ and $x - y \in [a - c, b - d]$. Is it always true that $xy \in [ac, bd]$? Or that $x/y \in [a/c, b/d]$?

5 TOPICS FOR FURTHER STUDY

One method that mathematicians have used to describe the real number system is to identify real numbers with splittings of the rational numbers into two nonempty complementary sets, one of which has no greatest element but contains every rational number less than any of its elements. Such splittings are called **Dedekind cuts**. In this scheme, every real number α is identified with a set of rationals that we would think of as having α for its supremum. All the properties of the real number system can be developed from this representation of real numbers; a complete exposition of this theory can be found in the classical text of Landau [2].

Open and closed intervals are the simplest examples of open and closed sets of numbers, two fundamental notions in topology. An open set of numbers is a set that contains an open interval about each of its elements. Every open set can be written as the union of a sequence of pairwise disjoint open intervals. Closed sets of numbers are characterized by the property of containing the supremum and infimum of each of their bounded, nonempty subsets. A closed set can also be described as a set whose complement is open. Closed sets can be much more complex than open sets, and there is no way to characterize them in terms of unions of closed intervals. For example, the Cantor set is a closed subset of $[0, 1]$ that contains no nondegenerate intervals, but it has so many points in it that no sequence of real numbers can possibly include all of them.

II

CONTINUOUS FUNCTIONS

As we try to make sense of our surroundings, we find that evolutionary change seems much simpler than abrupt change, because it is easier to recognize the underlying order. The same is true in mathematics; we need continuity to recognize order in potentially chaotic models.

1 FUNCTIONS IN MATHEMATICS

It seems that every modern calculus book discusses functions somewhere in the first chapter, often expecting the concept to be familiar from precalculus courses. Since functions are so commonplace, it's remarkable that our concept of function is a fairly recent one, unknown to the founders of calculus and quite different from what was expected by the first mathematicians to use abstract functions. The original idea was simply to extend algebraic notation in a way that lets us work with quantitative relationships abstractly, in much the same way that variables make it possible to look at arithmetic abstractly and develop general rules. Just as we often treat variables as unknowns and try to solve for them, we also sometimes treat functions as unknowns, and many important problems involve our trying to solve for them. But some of the things people found when they went looking for functions turned out to be very different from what they expected to find, and the need for careful definitions became apparent. Now we use functions to describe just about everything that can be considered a mathematical operation.

Mathematical operations don't just happen all by themselves; something must be operated on. And the operation must result in something, even if that something is a concept we associate with nothing, such as zero or the empty set. So we naturally associate mathematical operations with two sets, one for the input and one for the output. (Here we're using the notion of set in its most general sense—elements of the input or output might themselves be sets of numbers, for example.) The operation is called a function if the result of performing it depends only on the designated input, with a special emphasis on the word *designated*. For example, if the set for the input is a set of coins and the operation to be performed on a coin consists of determining its denomination, that's a function. However, if the operation consists of flipping the coin and noting whether it lands heads or tails, that's not a function because the outcome clearly depends on something besides the coin that was flipped. With this example in mind, here's a formal definition that spells out just when we can think of an operation as a function.

DEFINITION 1.1: Let A and B be sets, and let f be any process that can accept each element x of A as input and then produce an element of B as output. If the element in B that results is entirely determined by the element x selected from A, we say that f is a **function** from A to B. We indicate this symbolically as $f : A \to B$ and write $f(x)$ for the result of performing the operation f on the element x. We call the set A the **domain** of the function and the set B the **range** of the function. The subset of B given by $\{f(x) : x \in A\}$ is called the **image** of A under f.

The general concept of function is often illustrated concretely by examples where the domain or range consists of physical objects, such as the function whose domain is the set of all horses present in a given region at a given time and whose range is the set of integers, with the value of the function at each horse being its age in years. However, such examples fail to convey an important aspect of the way functions are used in mathematics. When the domain and range of a function are composed of mathematical objects (such as numbers) rather than physical objects, the rules change significantly. Physical objects often are one of a kind with a transitory existence; for example, if no two pieces of charcoal are exactly alike, then after you burn one to measure its value as an energy source you'll never be able to burn the same piece again. By virtue of their abstractness, mathematical objects are not subject to the same limitations. When you add 1 to 3 and get an answer of 4, you don't change 3 at all; the concept of 3 remains fixed. More importantly, a single mathematical object can appear in a variety of guises. We think of $2 + 1$, $6/2$, and $\sqrt{9}$ as exactly the same number as 3. So if f is a function and the number 3 is an element of its domain, then $f(3)$, $f(2+1)$, $f(6/2)$, and $f(\sqrt{9})$ must all refer the same element of the range of f. To a mathematician, writing $f : A \to B$ includes the requirement that when f operates on different representations of the same element of A, the results all represent the same element of B.

An understanding about when different expressions represent the same element of the domain or of the range is a necessary part of the definition of any particular function, even though such understandings are rarely stated explicitly. Such understandings explain why, for example, there is no function $f : \mathbf{R} \to \mathbf{R}$ with the property that $f(x^2) = x$ for all $x \in \mathbf{R}$; it's entirely possible to have $x_1^2 = x_2^2$ but $x_1 \neq x_2$. And we can't define $g : \mathbf{Q} \to \mathbf{Q}$ by $g\left(\frac{m}{n}\right) = \frac{m+2}{n}$, because the value of $\frac{m+2}{n}$ depends on both m and n, not just on the value of $\frac{m}{n}$ as an element of $\mathbf{Q}$. Subtle distinctions like this are an inescapable part of modern mathematics. After all, the notion that different things can sometimes be considered equal and that equal inputs should lead to equal outputs is fundamental. It's the guiding principle we follow when solving equations in algebra, for example. To a large extent, it's what separates mathematics from other things we might do with mathematical symbols. That's why almost every operation that's a part of mathematics can be considered a function.

In scientific work there's a somewhat different aspect to the problem of what constitutes a function. For example, suppose there are two parameters x and y that are used to describe some system that changes over time. We may make a number of simultaneous measurements of x and y, and then try to determine whether y can be considered a function of x. The mathematical criterion is simply that any two occasions producing equal x-values must also produce equal y-values, regardless of when those occasions take place. Usually the scientist can't examine all possible values of x and y, and the challenge is to develop a theory that explains why y should be a function of x when the observed values seem consistent with that assumption. That's a key part of the scientific method.

In formal mathematics, a function $f : A \to B$ consists of the domain A, the range B, and the rule used to determine $f(x)$ for arbitrary $x \in A$. But in calculus it's much more common to think of the function as simply the rule used to determine $f(x)$. For the most part, both the domain and range of the functions we work with are sets of real numbers; we call such functions **numerical functions**. Since our understanding about when two mathematical expressions represent the same real number generally doesn't change with the subsets of real numbers considered, we won't worry about specifying the domain or range of numerical functions unless there is a special need to do so. For instance, there is rarely any need to treat

$$f : \mathbf{R} \to \mathbf{R} \quad \text{defined by } f(x) = \frac{1}{x^2 + 1}$$

and

$$\tilde{f} : \mathbf{R} \to [0, 1] \quad \text{defined by } \tilde{f}(x) = \frac{1}{x^2 + 1}$$

as different functions, and we're not likely to do so. When we do use the notation $f : A \to B$ in reference to a numerical function, it's because we really do want to call attention to the domain and range; at some stage a property of one or both will be important.

In abstract mathematics, the **graph** of a function $f : A \to B$ is the set of all pairs of elements (x, y) with $x \in A$, $y \in B$, and $y = f(x)$. The definition of function guarantees that when (x_1, y_1) and (x_2, y_2) are in the graph of a single function f and $x_1 = x_2$, then $y_1 = y_2$ as well. In the case of numerical functions, both A and B are sets of real numbers and we usually identify the pair (x, y) with the point in the Cartesian plane having those coordinates, making the graph a set of points in the plane. In common usage, the graph of f usually refers to a pictorial representation of this set of points in the plane instead of to the set itself. Points (x_1, y_1) and (x_2, y_2) in the plane have $x_1 = x_2$ when the points are on the same vertical line, so the graph of a numerical function has the property that no vertical line can intersect it in more than one point. Graphs are a powerful tool for studying numerical functions because they present a whole range of values of a function simultaneously, rather than one value at a time. But the things we can do with graphs are limited, and an algebraic or analytic representation of a function is usually better suited to computations.

In addition to operating *with* numerical functions, we'll operate *on* them. Numerical functions are basic elements in a more abstract level of mathematics that we use to study the relationships between numerical variables, where we think of the relationships themselves as variables. We'll end this section with a description of some of the ways we combine functions.

Given two functions f and g with a common domain and both having ranges in $\mathbf{R}$, we can combine their values arithmetically to create new functions on the same domain. For example, we can always define $f + g$, $f - g$, and fg as real-valued functions on the same domain by simply performing the indicated operation on their values at each point in E. That is,

$$(f + g)(x) = f(x) + g(x),$$
$$(f - g)(x) = f(x) - g(x),$$
and $$(fg)(x) = f(x) g(x).$$

We can define constant multiples of f or g by the same process, and we can also define f/g if zero isn't a value of g.

We can also combine functions by the operation of composition, if their domains and ranges match up appropriately. The composite function $f \circ g$ is defined whenever the set of values of g is a subset of the domain of f. In that case the domain of $f \circ g$ is the domain of g and the range of $f \circ g$ is the range of f. All these conditions are obviously what we need to define the value $f \circ g$ at x to be $f(g(x))$. Whenever $g(x)$ represents an element of the domain of f, it makes perfectly good sense to perform the operation of f on it, and it produces the value $f(g(x))$.

EXERCISES

1. Find a numerical function f with domain $[1, \infty)$ such that

$$f\left(\sqrt{x^2 + 1}\right) = x^2 + 2x + 3\sqrt{x^2 + 1} \quad \text{for all } x \geq 0.$$

2. A moving point P in the coordinate plane is sometimes located in terms of its polar coordinates (r, θ); its Cartesian coordinates (x, y) are then

$$x = r \cos \theta \quad \text{and} \quad y = r \sin \theta.$$

While x, y, and r are considered to be functions of P, θ usually is not, because different values of θ correspond to the same point. Usually the value we use for θ depends on the route that was followed to arrive at P, not just the location of the point. With that in mind, explain why

$\sin 2\theta$ is considered a function of P, but $\sin \frac{1}{2}\theta$ is not. (The domain of the function does not include the origin.)

3. Suppose that f is a numerical function defined on $[a, b]$, and that s is a number such that $f(b) - f(a) \geq s(b - a)$. Use bisection and mathematical induction to prove that there is a nested sequence $\{[a_n, b_n]\}_{n=1}^{\infty}$ of closed intervals such that

$$b_n - a_n = 2^{-n}[b - a] \quad \text{and} \quad f(b_n) - f(a_n) \geq s(b_n - a_n)$$

for each interval $[a_n, b_n]$ in the sequence. Your proof should show how to construct the sequence in terms of the unspecified function f.

2 CONTINUITY OF NUMERICAL FUNCTIONS

The concept of function is so general that it is virtually impossible to visualize generic functions from $\mathbf{R}$ to $\mathbf{R}$. The operations used to evaluate $f(x)$ at different values of x can be entirely independent of each other. In principle there could be a different formula used at each x. Such functions have little practical value; we certainly can't use them to model physical processes. For example, if we can only measure temperatures to the nearest hundredth of a degree, how could we ever make use of a function of the temperature for which the difference between $f(98.6°)$ and $f(98.5999°)$ were significant? Unless approximate values for x can be used to find approximate values for $f(x)$, any physical system modeled by f will appear to behave so erratically that it's likely to be considered random rather than deterministic.

We deal with that problem when we impose the condition of continuity, a condition that we'll describe in several stages. For f a numerical function defined on all of $\mathbf{R}$, the condition is basically the same as the conditions we developed for doing exact arithmetic in $\mathbf{R}$ by means of approximations. That is, we say that f is continuous at the real number a if for every given $\varepsilon > 0$, there is a $\delta > 0$ such that $|f(x) - f(a)| < \varepsilon$ for all x with $|x - a| < \delta$. Study this statement carefully; it says a lot, and the logic of what it says is fairly complex. It has several distinct parts, and its meaning is wrapped up in the precise relationship between those parts. It's a statement about all of the values of f on each of a family of intervals centered at a. Trying to consider all the possible values of $f(x)$ as x ranges over a single interval is hard enough, and we have to consider what happens on many different intervals simultaneously. The task is simplified considerably when we can restrict our attention to an interval where the values of f either always increase or always decrease, but sometimes that isn't possible.

Deciding whether a given function is continuous at a given point looks like a difficult task, but the assumption of continuity is an extremely powerful one. In developing mathematical theories, knowledge of continuity is equivalent to possession of a machine that takes arbitrary positive epsilons and processes them into suitable positive deltas. That's the good news about continuity; the bad news is that proving a given function is continuous often amounts to assembling such a machine.

If our function is not defined on all of $\mathbf{R}$ but only on a subset of it, the condition we use to define continuity must be modified. For example, when $f : [a, b] \to \mathbf{R}$, there won't be any $\delta > 0$ for which we can say anything about the values of $f(x)$ at every $x \in (a - \delta, a + \delta)$. So we do the best we can, and require only that the condition on the values of $f(x)$ hold for all x in the intersection of $(a - \delta, a + \delta)$ and the domain of f. Since the condition always involves $f(a)$, we insist that a be a point in the domain of f before we attempt to spell out further conditions. Now we're ready to define continuity for numerical functions.

DEFINITION 2.1: For f a numerical function with domain E, we say that f is **continuous at the point** a provided that $a \in E$ and that for each given $\varepsilon > 0$ there is a $\delta > 0$ such that every $x \in (a - \delta, a + \delta) \cap E$ satisfies $|f(x) - f(a)| < \varepsilon$. Under the same conditions, we also say that a is a **point of continuity** for f. We say simply that f is a **continuous function** if every point in its domain is a point of continuity.

The beauty of this definition is that it is so thoroughly compatible with the things we've done so far and the things we'll want to do, once we get around to them. Obviously it's closely connected to our notion of approximation; if a numerical function f is continuous at a point a in its domain and we have a family of approximations to a involving only points where f is defined, then applying f to them produces a corresponding family of approximations to $f(a)$. It's often useful to reverse the process and think of $f(a)$ as approximating all the values of $f(x)$ at nearby points. When f is continuous, we can say a great deal about the properties of $f(x)$ over an entire interval in terms of its values at relatively few well-placed points. Arbitrary functions can be wild; by comparison, continuous functions are quite tame.

At the end of the last section, we mentioned some operations we used to combine functions. Such operations always preserve continuity; that's an extremely useful fact. It's so important that we'll state it as two formal theorems, even though the proofs are almost obvious.

THEOREM 2.1: *Let f and g be given numerical functions with domain E. Then each point of continuity for both f and g is a point of continuity for f + g, f − g, and fg. The same is true for f/g when zero is not in the image of E under g.*

  ☐ *Proof:* Note that continuity appears both in the hypotheses and in the conclusion. Effectively, the hypotheses give us some ε to δ machines that we can work with, and we link them together to build the machine called for in the conclusion. Our knowledge of approximate arithmetic lets us treat all three cases with the same argument. Consider the product fg, for example. Given $a \in E$ and $\varepsilon > 0$, we can guarantee that $|(fg)(x) − (fg)(a)| < \varepsilon$ by making sure that both $|f(x) − f(a)| < \delta_0$ and $|g(x) − g(a)| < \delta_0$ for some suitably chosen $\delta_0 > 0$. Since a is a point of continuity for f, we can choose $\delta_1 > 0$ such that

$$|f(x) − f(a)| < \delta_0 \quad \text{for all } x \in (a − \delta_1, a + \delta_1) \cap E.$$

Since a is also a point of continuity for g, we can also choose $\delta_2 > 0$ such that

$$|g(x) − g(a)| < \delta_0 \quad \text{for all } x \in (a − \delta_2, a + \delta_2) \cap E.$$

So we choose δ to be the smaller of the two positive numbers δ_1 and δ_2; then every $x \in (a − \delta, a + \delta) \cap E$ satisfies both conditions. Consequently, every $x \in (a − \delta, a + \delta) \cap E$ satisfies $|(fg)(x) − (fg)(a)| < \varepsilon$, and we've proved continuity at a.   ■

THEOREM 2.2: *Let f and g be numerical functions, with the range of g contained in the domain of f. If g is continuous at c and f is continuous at g(c), then f ∘ g is continuous at c.*

  ☐ *Proof:* As long as we understand what everything means, the proof of this theorem is straightforward. We should begin by noting that under the hypotheses, $f \circ g$ is a numerical function with the same domain as g. Let's call this set E and call $b = f(c)$. Thus we need to show that for any given $\varepsilon > 0$, there is a $\delta > 0$ such that every $x \in (c − \delta, c + \delta) \cap E$ satisfies

$$|f(g(x)) − f(b)| < \varepsilon.$$

Since we've assumed that f is continuous at b, we know that for our given ε there is a $\delta_0 > 0$ such that every y in the intersection of $(b − \delta_0, b + \delta_0)$ and the domain of f satisfies $|f(y) − f(b)| < \varepsilon$. By hypothesis, $g(x)$ is always in the domain of f, so all we need to do is to choose $\delta > 0$ such that every $x \in (c - \delta, c + \delta) \cap E$ satisfies $g(x) \in (b - \delta_0, b + \delta_0)$. The continuity of g at c is exactly the condition we need to assert that such a δ exists. ∎

By applying these theorems in various combinations, we can recognize the continuity of many functions defined by elementary formulas. In particular, if the values of a numerical function are determined by performing the same fixed, finite sequence of arithmetic operations at all points in the domain, then the function must be continuous. We won't try to write out a formal proof of this general principle; the difficulty in proving it has nothing whatever to do with the definition of continuity. Given any such function, proving that it is continuous becomes a straightforward process once we break it down to its simple parts. Rather, the difficulty in proving the general principle is in characterizing those functions in such a way that the argument given clearly applies to all of them.

When we're trying to establish the continuity of a single function that we don't immediately recognize as continuous, we often study the rule defining the function and try to bound $|f(x) - f(a)|$ in terms of $|x - a|$. The general principle involved—which we often use with no conscious effort—deserves to be stated as a theorem even though its proof is extremely simple.

THEOREM 2.3: *Suppose that f is a numerical function with domain E, $a \in E$, and g is another numerical function such that*

$$|f(x) - f(a)| \leq g(x - a) \quad \text{for all } x \in E.$$

If g is continuous at 0 with $g(0) = 0$, then f is continuous at a.

□ *Proof:* All we have to do is recognize that when $\varepsilon > 0$ is given, the δ given by the continuity of g at 0 can serve as the δ needed for continuity of f at a. ∎

The definitions of simple functions often incorporate operations other than addition, subtraction, multiplication, and division, such as deciding such as whether $x > a$ is true. That decision and others like it can introduce discontinuities; every interval $(a - \delta, a + \delta)$ contains both x for which $x > a$ is true and x for which it is false. Such decisions can present practical difficulties when we attempt to evaluate the function at a value determined only approximately, but they offer no logical problems and are frequently used in mathematical theories.

According to what we said earlier, $1/x$ should be considered a continuous function. Its domain is the set of all nonzero real numbers and it's continuous at each point in its domain. But in common usage, we also say that $1/x$ is discontinuous at $x = 0$, and it seems contradictory to talk about the discontinuities of continuous functions. The reason for the apparent paradox is that we haven't yet given a precise definition of discontinuity. In common usage, the notions of discontinuity at a point and continuity at a point aren't quite exact opposites. When we're considering functions that aren't defined on all of $\mathbf{R}$, we sometimes consider points outside the domain to be discontinuities.

The reader should already recognize that a numerical function f with domain E is discontinuous at a point $a \in E$ if for some $\varepsilon > 0$ every interval of the form $(a - \delta, a + \delta)$ will contain a point $x \in E$ with $|f(x) - f(a)| \geq \varepsilon$, no matter how small we choose $\delta > 0$. Indeed, to prove a given function f is discontinuous at a given point a in its domain, we usually must produce such an ε. But we also consider f to be discontinuous where points are missing from its domain. That is, f is discontinuous at a point b not in its domain E if every interval $(b - \delta, b + \delta)$ contains points of E. The function f defined by $f(x) = 1/x$ has this sort of a discontinuity at 0. While 0 isn't in the domain of f, every interval $(-\delta, \delta)$ contains points that are. While we consider $1/\sqrt{x}$ to be discontinuous at 0, we don't consider it to be discontinuous at -1 because an interval $(-1 - \delta, -1 + \delta)$ need not contain any points in the domain of $1/\sqrt{x}$.

According to our definitions, a function is either continuous or discontinuous at any given point in the domain. While a point outside the domain can never be a point of continuity, it may or may not be a discontinuity; that's determined by the relationship of the point to the domain. A continuous function can indeed have discontinuities, but only at points outside its domain.

EXERCISES

4. Use the definition of continuity to prove that the numerical function f defined by $f(x) = 2x$ for all $x \in \mathbf{R}$ is continuous.

5. Use the definition of continuity to prove that if f is continuous at a point a and $f(a) > 0$, then there is an open interval $(a - \delta, a + \delta)$ containing no x with $f(x) \leq 0$.

6. Let g be the numerical function whose value at each $x \in \mathbf{R}$ is the decimal value of x, rounded to the nearest hundredth. Show that g has discontinuities by producing an $a \in \mathbf{R}$ and an $\varepsilon > 0$ for which no corresponding $\delta > 0$ exists. How small does ε need to be in this case?

7. Why is the function g in the previous exercise continuous on the interval $(-.005, .005)$?

8. Suppose that f is a numerical function with domain E, and that a is a point of continuity of f. Prove that there is an open interval I containing a for which the set of values $\{f(x) : x \in I \cap E\}$ has a supremum and an infimum.

9. Prove that if f is an increasing function on (a, b) and the set of values $\{f(x) : a < x < b\}$ is also an interval, then f must be continuous on (a, b). (The conclusion remains true when (a, b) is replaced by an arbitrary interval, but more cases must be considered to prove it.)

3 THE INTERMEDIATE VALUE THEOREM

Functions defined and continuous on an interval are often thought of as having graphs that form continuous paths. This evokes the image of a graph that can be sketched without lifting the pencil used to draw it. That's thought to be the reason we use the word *continuous* to describe such functions, even though that's not the mathematical meaning of the word. The theorem below, universally known as the **intermediate value theorem**, is closely related to this naive description of continuity.

THEOREM 3.1: *Let f be a numerical function that is continuous on $[a, b]$. Then for any number r between $f(a)$ and $f(b)$, there is a real number c in $[a, b]$ with $f(c) = r$.*

The graphical interpretation of this theorem is so striking that students instinctively accept it as true. This interpretation says that when the horizontal line $y = r$ separates the points $(a, f(a))$ and $(b, f(b))$, the graph of f must cross the line at some point whose x-coordinate is between a and b. Obviously the graph must also cross any vertical line that separates the points, but we don't need a theorem to tell us that. Since we know the x-coordinate we can just evaluate f to locate the point on a given vertical line. The intermediate value theorem deals with something that can be a real problem: finding a value of x that satisfies the equation $f(x) = r$. We should not dismiss the theorem as obviously true on geometrical grounds, because if the graph had any gaps it could easily appear on both sides of a line without actually crossing it anywhere, and the actual definition of continuous function doesn't explicitly say anything about the presence or absence of gaps in the graph. Continuous functions aren't necessarily as nice as the smooth curves we use to visualize their graphs, so in our proof we'll have to be careful that we don't assume something that isn't actually part of the hypotheses.

Before we prove the theorem, let's consider an example and see how we might deal with it. Let f be the function defined by $f(x) = x^5 + x$. It's continuous on all of $\mathbf{R}$ but we'll focus our attention on $[0, 1]$. Obviously $f(0) = 0$ and $f(1) = 2$, so a simple application of the intermediate value theorem says there must be an $x \in [0, 1]$ with $x^5 + x = 1$. Algebra doesn't offer any method for solving this equation to find x, but with the help of a calculator we can get good results with a trial-and-error procedure.

An examination of the formula defining f shows us that increasing x always increases $f(x)$, so it's easy to bracket the solution: x is too big if $f(x) > 1$ and too small if $f(x) < 1$. We first try the midpoint of $[0, 1]$. Since $f(0.5) < 1$ we narrow our search for x to $[0.5, 1]$. The midpoint of this interval is 0.75, and since $f(0.75) < 1$, we narrow our search for x to $[0.75, 1]$. At the next midpoint we find $f(0.875) > 1$, so x must be in $[0.75, 0.875]$. This process produces a nested sequence of closed intervals, and the value of x we want must be the one number common to all of them. Watching the trial values of f approach 1 provides strong evidence that x really will have $f(x) = 1$.

What if we try to adapt this argument to an unspecified function? The idea of finding c by producing a nested sequence of closed intervals certainly looks appealing, but we'll need some modifications. In particular, an unspecified function might have either $f(a) < r < f(b)$ or $f(b) < r < f(a)$ and still satisfy the hypotheses of the theorem, so we have to deal with both possibilities. But a much greater problem is that we won't actually look at any numerical values of f, so we'll need some other consideration to explain why the c we find really will satisfy $f(c) = r$. That's where the precise definition of continuity will come into play.

☐ *Proof:* For the general case, we first produce a nested sequence of intervals

$$\{[a_n, b_n]\}_{n=1}^{\infty},$$

starting with $a_1 = a$ and $b_1 = b$. All of the different cases with r between $f(a_n)$ and $f(b_n)$ can be summarized in a single inequality, one that includes the cases $f(a_n) = r$ or $f(b_n) = r$ as well:

$$P_n = [f(a_n) - r][f(b_n) - r] \le 0. \tag{2.1}$$

That inequality will also help us prove that the c we find will have $f(c) = r$. For $n = 1$ we have $P_1 \le 0$ by hypothesis, and we produce our sequence inductively in a way that makes sure $P_n \le 0$ for all n. Let's assume that $[a_k, b_k]$ has been chosen with $P_k \le 0$, then call m_k the midpoint of $[a_k, b_k]$. We'll choose $[a_{k+1}, b_{k+1}] = [a_k, m_k]$ unless $[f(a_k) - r][f(m_k) - r] > 0$, and in that case we'll choose $[a_{k+1}, b_{k+1}] = [m_k, b_k]$ instead. This procedure guarantees that $P_{k+1} \le 0$. Note that

$[f(a_k) - r][f(m_k) - r] > 0$ implies $[f(m_k) - r][f(b_k) - r] \leq 0$ since their product is $[f(m_k) - r]^2 P_k$ and we assumed that $P_k \leq 0$.

Now let c be the one number common to all these intervals; it's certainly a point in our original interval $[a, b]$. To finish the proof we must show that $f(c) = r$, and we do it by showing that $f(c) \neq r$ requires f to have a discontinuity at c. Our hypotheses include the continuity of f on all of $[a, b]$, so we can then conclude that $f(c) = r$.

For $f(c) \neq r$, every number close enough to $[f(c) - r]^2$ must be positive, so our rules for approximate arithmetic say there is an $\varepsilon > 0$ such that

$$(u - r)(v - r) > 0 \quad \text{for all } u, v \in (f(c) - \varepsilon, f(c) + \varepsilon).$$

But no matter how small we pick $\delta > 0$, we can find n large enough that both a_n and b_n are in $(c - \delta, c + \delta) \cap [a, b]$, and since $P_n \leq 0$ we know that one or both of the numbers $f(a_n)$, $f(b_n)$ must be outside the interval $(f(c) - \varepsilon, f(c) + \varepsilon)$. Thus if $f(c) \neq r$ then f must have a discontinuity at c, and that completes the proof. ∎

We may note that if r is between $f(a)$ and $f(b)$ but doesn't equal either of them, then of course the c we find with $f(c) = r$ can't be either a or b, and therefore satisfies $a < c < b$. Consequently, there's no reason to worry about whether the hypothesis that r is between $f(a)$ and $f(b)$ includes the case of equality; the versions of the theorem with either interpretation are clearly equivalent.

We should note that the intermediate value theorem gives only a property that continuous functions have, not a definition of continuity. The standard example of a discontinuous function that satisfies the conclusion of the intermediate value theorem is

$$f(x) = \begin{cases} \sin(1/x), & x > 0 \\ 0, & x \leq 0. \end{cases}$$

In every interval $(0, \delta)$, f takes on every value in $[-1, 1]$, and that makes it easy to see that intermediate values will always exist even though f is not continuous at 0. An even more striking example is given in the book by R. P. Boas [1]; he defines a function f such that for every open subinterval (a, b) of $[0, 1]$, the set of values $\{f(x) : a < x < b\}$ is the entire interval $[0, 1]$.

EXERCISES

10. Suppose that f is a continuous numerical function on an interval I and that V is the set of values $\{f(x) : x \in I\}$. Prove that V is an interval. (It makes no difference whether the interval I is open or closed, bounded or unbounded. Don't try to determine the type of intervals involved.)

11. Suppose that $f : \mathbf{R} \to \{0, 1\}$ and that I is an interval on which f has both values. Prove that f has a discontinuity in I.

12. Suppose that f is a continuous function on an interval I, $a \in I$, we've been given an $\varepsilon > 0$, and we're actually trying to calculate a positive δ such that every $x \in (a - \delta, a + \delta) \cap I$ satisfies $|f(x) - f(a)| < \varepsilon$. Show that δ is suitable if $(a - \delta, a + \delta)$ contains no x with $f(x) = f(a) \pm \varepsilon$.

13. For $f(x) = 1/x$ and $a > 0$, find the largest possible δ with the property that every $x \in (a - \delta, a + \delta)$ satisfies $|1/x - 1/a| < \varepsilon$ by solving

$$1/(a + \delta) = 1/a - \varepsilon \quad \text{and/or} \quad 1/(a - \delta) = 1/a + \varepsilon.$$

Finding δ is easy; explain how you know it will have the desired properties.

14. If f and g are continuous numerical functions on $[a, b]$ with $f(a) < g(a)$ but $f(b) > g(b)$, why must there be an $x \in (a, b)$ with $f(x) = g(x)$?

4   MORE WAYS TO FORM CONTINUOUS FUNCTIONS

In addition to the algebraic formulas we use to define specific functions, we sometimes consider functions for which the method used to determine $f(x)$ depends on whether x is in one or more specially designated sets. While purely algebraic formulas generally produce only functions that are continuous at all points in their domain, a formula that incorporates a decision as to whether or not a value of the variable has a specific property can easily lead to discontinuities. You may have noticed that such decision rules have a way of appearing whenever anyone gives an example of a function with a discontinuity at one or more points in its domain. However, decision rules don't automatically lead to discontinuities; let's look at some examples of what can happen.

The simplest continuous function involving a decision rule is the absolute value function,

$$|x| = \begin{cases} x, & x \geq 0 \\ -x, & x < 0. \end{cases}$$

The absolute value function is continuous on all of $\mathbf{R}$; indeed, for any $\varepsilon > 0$ we know every $x \in (a - \varepsilon, a + \varepsilon)$ satisfies $||x| - |a|| < \varepsilon$. Consequently, when f is any numerical function we see that the function $|f|$ defined by

$$|f|(x) = |f(x)|$$

is continuous at every point where f is continuous. To satisfy the definition of continuity, when $\varepsilon > 0$ is given we can use any δ for $|f|$ that would work for f.

Another way to look at the absolute value function is that evaluating it amounts to choosing the larger of two numbers whose average is 0. We can also use absolute values to select the greater or lesser of any two numbers a and b. Since their average is halfway between them, both are at the distance $\frac{1}{2}|a - b|$ from $\frac{1}{2}(a + b)$, and so

$$\max\{a, b\} = \frac{1}{2}(a + b) + \frac{1}{2}|a - b|,$$

$$\min\{a, b\} = \frac{1}{2}(a + b) - \frac{1}{2}|a - b|.$$

That makes it easy to recognize that whenever f and g are continuous functions on the same domain, selecting the larger of the two values $f(x)$ and $g(x)$ at each point x defines another continuous function, and so does selecting the smaller value at each point, because both $\frac{1}{2}(f + g)$ and $\frac{1}{2}|f - g|$ define continuous functions.

Our formula for the absolute value function is the most familiar example of a useful technique: we take two or more functions with different domains and patch them together to define a single function on the union of the two domains. That is, we start with $f : A \to \mathbf{R}$ and $g : B \to \mathbf{R}$, and define $h : A \cup B \to \mathbf{R}$ so that $h(a) = f(a)$ for $a \in A$ and $h(b) = g(b)$ for $b \in B$. That works just fine when A and B have no elements in common, but it may not work when they do. For $x \in A \cap B$, should $h(x)$ be $f(x)$ or should it be $g(x)$? If the rule we use to decide depends on anything but the value of x, it would seem to contradict the definition of function. But in fact it may not; the definition requires that the **value** we assign to $h(x)$ depend only on the value of x; it says nothing about the **process** we use to determine the value of $h(x)$. In particular, if $f(x) = g(x)$, then it doesn't really matter whether we say $h(x) = f(x)$ or $h(x) = g(x)$. That is, as long as $f(x) = g(x)$ for all $x \in A \cap B$, we have no problem defining $h : A \cup B \to \mathbf{R}$ in terms of f and g. We call h the **common extension** of f and g to the union of their domains.

What about the continuity of the common extension of two functions? There are some simple implications here that we'll state in a formal theorem. Its proof is quite easy, chiefly because it fails to address the difficult cases.

THEOREM 4.1: *For A and B subsets of* **R**, *suppose that we have two functions* $f : A \to$ **R** *and* $g : B \to$ **R** *whose values agree on* $A \cap B$. *Then their common extension is continuous at each point where both f and g are continuous, and any discontinuity of their common extension must be a discontinuity for f or for g, if not for both.*

☐ *Proof:* Since every point in the domain of a function is either a discontinuity or a point of continuity, the first part of the theorem is a logical consequence of the second so we'll just prove the second part. To fix the notation, let's call h the extension of f and g to $A \cup B$.

We begin by considering the possibilities at a point $a \notin A \cup B$. Either f has a discontinuity at a or there is a $\delta_1 > 0$ such that $(a - \delta_1, a + \delta_1)$ contains no points of A, and either g has a discontinuity at a or there is a $\delta_2 > 0$ such that $(a - \delta_2, a + \delta_2)$ contains no points of B. So when neither f nor g has a discontinuity at a, we can take δ to be the smaller of δ_1 and δ_2 to get an interval $(a - \delta, a + \delta)$ that doesn't contain any points of $A \cup B$. This shows that h doesn't have a discontinuity at a, and we've taken care of the case $a \notin A \cup B$.

Next let's suppose that $a \in A$ and that h has a discontinuity at a. There is nothing to prove if f has a discontinuity at a, so we might as well assume that f is continuous at a. Since a is in the domain of h, there must be an $\varepsilon > 0$ such that every interval $(a - \delta, a + \delta)$ contains at least one $x \in A \cup B$ with

$$|h(x) - h(a)| \geq \varepsilon.$$

But f is continuous at a, so when δ is small enough we know that every x in $A \cap (a - \delta, a + \delta)$ satisfies $|f(x) - f(a)| < \varepsilon$. Since f and h agree on A, any $x \in (a - \delta, a + \delta)$ with $|h(x) - h(a)| \geq \varepsilon$ must be in B, not A. Hence every open interval centered at a must contain $x \in B$ with

$$|g(x) - h(a)| \geq \varepsilon.$$

Since $h(a)$ agrees with $g(a)$ when $a \in B$, it follows that g has a discontinuity at a, whether or not $a \in B$.

The case $a \in B$ can be argued similarly; we just interchange the roles of f and g. That concludes the proof. ∎

We routinely use this theorem to check the continuity of a class of custom-built functions commonly known as **spline functions**. We take continuous functions defined on adjacent intervals and patch them together.

If the intervals are closed intervals and the values of the functions match at the endpoints, then their common extension is necessarily continuous.

It's entirely possible for the common extension of two functions to be continuous at a point where one or the other has a discontinuity, as long as the point isn't in the domain of both. For example, if $f(x) = x^2/x$ for $x \neq 0$ and $g(x) = x$ for $x \geq 0$, then their common extension is continuous everywhere, even though f has a discontinuity at the origin. Of course, one could argue that such a discontinuity shouldn't even count. Such discontinuities are called **removable**. Here's a formal definition of that concept.

DEFINITION 4.1: A function f has a **removable discontinuity** at the point a if f has a discontinuity there, but there is a function g that is continuous at a and that agrees with f everywhere else in the domain of f.

In other words, a removable discontinuity can be transformed into a point of continuity by either defining or correcting the definition of the function at the point. No points need to be deleted from the domain and the values of the function need not be changed at any points except the one where the removable discontinuity occurs. We call such a modification removing the discontinuity; it can be as simple as rewriting $\frac{x^2-1}{x-1}$ as $x+1$ to remove the discontinuity at $x = 1$. On the other hand, if f has a discontinuity at a and there is an $\varepsilon > 0$ such that every interval $(a - \delta, a + \delta)$ contains a pair of points $x, y \neq a$ with $|f(x) - f(y)| \geq 2\varepsilon$, then the discontinuity is certainly not removable. However $f(a)$ is defined, one or both of the numbers $|f(x) - f(a)|$ and $|f(y) - f(a)|$ must be at least ε. Basic calculus includes many problems that can be solved by recognizing discontinuities as removable and then removing them. The next chapter deals with the abstract theory behind removing discontinuities.

The reader should have noticed that the precise domain of the function being considered has a great deal to do with answering questions of continuity. On many occasions we have no interest in the values of a numerical function at all points of its domain, but only at points in some subset. In formal mathematics, it is then common to introduce a new function, called the **restriction** of the original function to the set of interest. That is, when A is a subset of the domain of f, the restriction of f to A keeps the same values as f at each $x \in A$, but we think of A as its entire domain. For example, if we define f on all of **R** by the simple rule

$$f(x) = \begin{cases} 0, & x < 0 \\ 1, & x \geq 0, \end{cases}$$

then f has a discontinuity at the origin. However, the restriction of f to $[0, \infty)$ is continuous; the origin becomes a point of continuity when we change the domain. In calculus we are often interested only in the values of a function on a particular interval, but since we seldom specify the domain of the functions we work with, we'll adopt the following convention.

DEFINITION 4.2: We say that the numerical function f is **continuous on the interval I** if the domain of f includes I and the restriction of f to I is a continuous function.

EXERCISES

15. Find a continuous numerical function on $[0, 2]$ whose graph includes the points $(0, 2)$, $(1, 0)$, and $(2, 1)$ and follows straight-line paths between them.

16. Use the definition of continuity at a point to prove directly that the common extension of f and g is continuous at each point where both are continuous.

17. Write out a proof that if the common extension of f and g has a discontinuity at a point where g is continuous, then f must have a discontinuity there. (This was the last case considered in the theorem proved in this section.)

18. The function f defined by $f(x) = x/|x|$ has a discontinuity at the origin. Either show why this discontinuity is not removable, or find a way to remove it.

19. Find the discontinuities of the function defined by the formula

$$\frac{x^2 + x - 6}{x^3 - 3x^2 + 2x}.$$

Which are removable?

5 EXTREME VALUES

Finding the greatest or least value of a function is an inherently difficult procedure to perform analytically because it involves considering all the values simultaneously, while analytic methods generally consider only one value at a time. Looking at the values one at a time won't even let you determine whether the set of values has upper or lower bounds, and if the set of values does have a supremum or an infimum there is still the question of determining whether they are actually values of the function. With many functions, there are ways to get enough information about their graphs to determine the extreme values, but in fact there is no general method for doing so. Consequently, in this section we will often be able only to prove that certain things exist without necessarily having a workable procedure for finding them. That stands in stark contrast to our earlier work of proving the existence of intermediate values by creating a procedure for producing them.

Let's consider the problem of finding the minimum value of a numerical function f on a closed interval $[a, b]$; the problem of finding the maximum is not very different. First assume that the minimum occurs at a point $c \in [a, b]$. If we knew the value of c, it would be easy enough to produce a nested sequence $\{[a_n, b_n]\}_{n=1}^{\infty}$ of closed subintervals of $[a, b]$ with c the one number common to all of them. We could take $[a_1, b_1] = [a, b]$ and when $[a_n, b_n]$ has been chosen with $c \in [a_n, b_n]$, we then define $m_n = \frac{1}{2}(a_n + b_n)$ and take

$$[a_{n+1}, b_{n+1}] = \left\{ \begin{array}{ll} [a_n, m_n], & c \leq m_n \\ {[m_n, b_n]}, & c > m_n. \end{array} \right.$$

We wouldn't really need to do any of this if we knew c, but the idea suggests an interesting possibility: can such a sequence be used to prove the existence of c?

It is indeed possible, but if we are going to use the sequence to prove the existence of c then we must describe the sequence in a way that makes no direct reference to either c or its existence. Instead of asking that each interval $[a_n, b_n]$ contain a single number c with the property that $f(c) \leq f(x)$ for every $x \in [a, b]$, we ask that for each $x \in [a, b]$ there be an element $x_n \in [a_n, b_n]$ with $f(x_n) \leq f(x)$. Such a sequence of intervals must exist for any numerical function on any closed interval $[a, b]$, and when f is continuous on $[a, b]$ its minimum value must occur at the one point common to all of the intervals in the sequence. We'll show how this works as we prove the theorem below. It seems both natural and convenient to refer to the theorem as the **extreme value theorem**, but this name is not in common use.

THEOREM 5.1: *Every numerical function that is continuous on a closed interval has both a greatest and a least value on that interval.*

□ *Proof:* We'll only show why f has a minimum value; the maximum value could always be found in terms of the minimum of $-f$. We use induction to prove the existence of a sequence of intervals like we described above. Getting the first interval in the sequence is easy: $[a_1, b_1] = [a, b]$. Given $x \in [a, b]$ we can choose $x_1 = x \in [a_1, b_1]$ to get $f(x_1) \leq f(x)$, so $[a_1, b_1]$ has the property we're looking for. To prove the existence of the entire sequence, we assume that $[a_n, b_n]$ is a closed subinterval of $[a, b]$ with length $2^{1-n} (b - a)$ and that for each $x \in [a, b]$ there is an $x_n \in [a_n, b_n]$ with $f(x_n) \le f(x)$. Then we show that $[a_n, b_n]$ must contain a subinterval $[a_{n+1}, b_{n+1}]$ with length $2^{-n} (b - a)$ and having the property that for each $x \in [a, b]$, some $x_{n+1} \in [a_{n+1}, b_{n+1}]$ satisfies $f(x_{n+1}) \le f(x)$.

As usual, we call m_n the midpoint of $[a_n, b_n]$. The next interval can be $[a_n, m_n]$ unless some $x^* \in [a, b]$ satisfies $f(x^*) < f(x)$ for all $x \in [a_n, m_n]$, and it can be $[m_n, b_n]$ unless some $x^{**} \in [a, b]$ satisfies $f(x^{**}) < f(x)$ for all $x \in [m_n, b_n]$. If such x^* and x^{**} both existed, the smaller of the values $f(x^*)$ and $f(x^{**})$ would be strictly less than every value of f on $[a_n, b_n]$, violating our assumption about $[a_n, b_n]$. So we see that a suitable $[a_{n+1}, b_{n+1}]$ must exist.

Now we use continuity to prove that f must have its minimum value on $[a, b]$ at the one point c common to all the intervals in our nested sequence. Obviously, $f(c)$ is a value of f on $[a, b]$. We prove that it is the minimum value by showing that no $y < f(c)$ can be a value of $f(x)$ with $x \in [a, b]$. Given $y < f(c)$, we choose $\varepsilon > 0$ with $y < f(c) - \varepsilon$. By the definition of continuity at c, there must be a $\delta > 0$ such that every $x \in (c - \delta, c + \delta) \cap [a, b]$ satisfies $|f(x) - f(c)| < \varepsilon$ and hence $f(x) > y$. But when n is large enough to make $b_n - a_n < \delta$ we must have $[a_n, b_n] \subset (c - \delta, c + \delta)$. No $x_n \in [a_n, b_n]$ could ever satisfy $f(x_n) \le y$. Thus y can't be $f(x)$ for any choice of $x \in [a, b]$, and that proves the theorem. ∎

Now let's turn our attention to another sort of extreme value problem. If we assume that f is a continuous numerical function on $[a, b]$, then for $\varepsilon > 0$ and each $c \in [a, b]$, we know that there must be a $\delta > 0$ such that every $x \in (c - \delta, c + \delta) \cap [a, b]$ satisfies $|f(x) - f(c)| < \varepsilon$. We may also note that the smaller the slope of the graph near $(c, f(c))$, the larger δ can be. Conversely, where the slope is large we may need smaller values for δ. But there is no bound on how steep the slope of the graph might get, so it appears that for a given ε we might need to use arbitrarily small positive values for δ. The theorem below shows that's not the case when we restrict our attention to a closed interval on which f is continuous; for each $\varepsilon > 0$, a single value of δ can be used at all points in the interval.

THEOREM 5.2: *Let f be a numerical function that is continuous on $[a, b]$. Then given any $\varepsilon > 0$, there is a $\delta > 0$ such that $|f(x) - f(y)| < \varepsilon$ for every x and y in $[a, b]$ with $|x - y| < \delta$.*

The conclusion of this theorem involves a condition that is so useful that we give it a special name; we say that the function is **uniformly**

continuous. Explicitly, when f is a numerical function with domain E, we say that f is uniformly continuous if for each $\varepsilon > 0$ there is a $\delta > 0$ such that every x and y in E with $|x - y| < \delta$ satisfy $|f(x) - f(y)| < \varepsilon$.

  □ *Proof:* We begin our proof by restating the conclusion: there is a number $\delta > 0$ such that every subinterval $[y, z]$ of $[a, b]$ with $|f(y) - f(z)| \geq \varepsilon$ has length at least δ. Let's give a name to the subintervals we have to watch out for: call

$$\mathcal{E} = \{[y, z] \subset [a, b] : |f(y) - f(z)| \geq \varepsilon\}.$$

Any δ will work if $\mathcal{E} = \emptyset$, so we may as well assume that $\mathcal{E} \neq \emptyset$. In that case, we must prove there is a positive number δ that is less than or equal to the length of every interval in $\mathcal{E}$.

For $\mathcal{E} \neq \emptyset$, we can define a numerical function Δ on $[a, b]$ by the formula

$$\Delta(x) = \inf\{r > 0 : (x - r, x + r) \text{ has a subinterval in } \mathcal{E}\}.$$

Clearly, $0 \leq \Delta(x) \leq b - a$ for each $x \in [a, b]$. For any $[y, z] \in \mathcal{E}$, we also see that $\Delta(x) \leq y - z$ at each $x \in [y, z]$, so the δ we need can be any number such that $\delta \leq \Delta(x)$ for all $x \in [a, b]$. So if we prove that Δ is continuous on $[a, b]$ and all its values are positive, we can appeal to the extreme value theorem and call δ the minimum value of Δ.

In fact, Δ is continuous on $[a, b]$ whether or not f is continuous. To prove it, we go back to the definition of Δ and find a way to compare its values at two points x and x'. For any $r > 0$, we have

$$(x - r, x + r) \subset \left(x' - |x - x'| - r, x' + |x - x'| + r\right), \qquad (2.2)$$

so if the interval on the left-hand side of (2.2) has a subinterval in $\mathcal{E}$ then so does the one on the right. Consequently,

$$\Delta(x') \leq |x - x'| + \Delta(x),$$

and since the roles of x and x' can be switched, we find

$$\left|\Delta(x) - \Delta(x')\right| \leq |x' - x|.$$

That actually shows that Δ is uniformly continuous, but that isn't important here. All we wanted is simple continuity.

When f is continuous on $[a, b]$, then $\Delta(c)$ is positive at each c. In fact, $\Delta(c) \geq r$ whenever r is small enough that every $x \in (c - r, c + r) \cap [a, b]$ satisfies $|f(x) - f(c)| < \varepsilon/2$. After all,

$$|f(y) - f(z)| \leq |f(y) - f(c)| + |f(z) - f(c)|,$$

so one or both of the endpoints of any interval in $\mathcal{E}$ must fall outside the interval $(c - r, c + r)$. Hence the minimum value of Δ is positive, and that completes the proof. ∎

Here's an important application of the last theorem. Suppose that we're interested in plotting the graph of a numerical function f that is continuous on $[a, b]$, and we want the y-coordinates of our graph to be accurate to within ε at all points of that interval. If we find the number δ given by the theorem, we can then pick an integer n with $n\delta > b - a$. Then we call

$$x_k = a + \frac{k}{n}(b - a) \quad \text{for } k = 0, 1, \ldots, n.$$

For $x \in [x_{k-1}, x_k]$, we'll have $|x_{k-1} - x| < \delta$ and $|x_k - x| < \delta$. So both $f(x_{k-1})$ and $f(x_k)$ will be in $(f(x) - \varepsilon, f(x) + \varepsilon)$. Every number between $f(x_{k-1})$ and $f(x_k)$ must also be in $(f(x) - \varepsilon, f(x) + \varepsilon)$. So the broken-line graph we obtain by successively connecting the points $\{(x_k, f(x_k))\}_{k=0}^{n}$ will have the desired accuracy.

EXERCISES

20. Prove that if f is a continuous numerical function on a closed interval I, then the set of values $\{f(x) : x \in I\}$ is itself a closed interval.

21. Modify the argument given in the proof of the extreme value theorem to prove the existence of a greatest value.

22. In the case of $\sqrt{x}$ on $[0, 1]$ with ε given in $(0, 1)$, show that ε^2 is the largest value of δ satisfying the conclusion of Theorem 5.2.

23. If a numerical function f is increasing and continuous on $[a, b]$, the intermediate value theorem can be used to find the δ in Theorem 5.2. Choose an integer n larger than $2[f(b) - f(a)]/\varepsilon$, and explain why for $k = 0, 1, 2, \ldots, n$ it is possible to find x_k in $[a, b]$ such that

$$f(x_k) = f(a) + \frac{k}{n}[f(b) - f(a)].$$

Explain why the least of the numbers $\{x_k - x_{k-1} : 1 \le k \le n\}$ is a suitable δ.

III

LIMITS

What we can actually do in mathematics is an incredibly small part of what we would like to do. Operations we can perform on numbers don't really go much beyond the addition, subtraction, and multiplication of whole numbers; we are limited in what we do by a fundamental restriction. Namely, we must be able to finish the job. If we can't finish it, then it is not something we can actually do, even if we're capable of performing each step the job requires. This is the sort of limitation we run into in trying to calculate $\sqrt{2}$, for example. But the mere fact that we humans can't complete a process doesn't stop us from studying it, theorizing about it, and trying to find what would result if it were somehow completed. By studying the theoretical result of processes we can't complete, we've made many important discoveries. Sometimes the obstacle is more imagined than real, and while it blocks an approach it may leave the goal itself accessible.

Calculus provides us with a systematic way to transcend the limitations on the calculations we can complete, and that makes it one of the greatest intellectual achievements of our civilization. The one concept that makes all this possible is the notion of a **limit**, a word we use to describe some operations that appear to involve a step just beyond what we can complete. That is, while we may never be able to complete all the steps required to calculate limits, we can approximate them arbitrarily closely by quantities that we do have methods for computing. We discover that it is quite possible to work with limits, and it is often even a simple matter to find exact relationships between quantities defined in terms of them.

Many different sorts of limits are used in calculus. They amount to simple variations on a basic theme, and most of them come quite readily once a single basic notion of limit has been mastered. For that reason, we'll begin our study of limits with the one that is easiest to understand: the limit of a sequence of numbers.

I SEQUENCES AND LIMITS

Let $\{x_n\}_{n=1}^{\infty}$ be a sequence of real numbers. That is, for each natural number n there is a real number x_n defined. We do not concern ourselves with how the process defining the sequence works. But we are concerned with properties of the sequence as a whole, not just with individual terms. In principle there is no obstacle to examining the beginning terms of the sequence one at a time, no matter how many we have to look at; it's dealing with the rest of the terms that's a challenge. We call a set $\{x_n : n \geq m\}$ a **tail** of the sequence, and this is the part that holds our interest for now. If we can find some way to deal with a tail all at once, then studying properties of the entire sequence presents no insurmountable difficulty.

Convergent sequences have such a property; they have tails in which all the terms are nearly equal. Let's explore this notion further before we attempt to formulate a precise definition. Assuming that $\{x_n\}_{n=1}^{\infty}$ is a bounded sequence of real numbers, we can use the tails to define a sequence of intervals $\{[a_m, b_m]\}_{m=1}^{\infty}$ by setting

$$a_m = \inf\{x_n : n \geq m\} \quad \text{and} \quad b_m = \sup\{x_n : n \geq m\}.$$

Recall that we don't call $[a_m, b_m]$ an interval unless $a_m \leq b_m$, and we know that's true because $a_m \leq x_m \leq b_m$ for each m. The sequence of intervals we've defined is actually a nested sequence of closed intervals. They're nested because a_m and b_m are bounds for $\{x_n : n \geq m+1\}$ as well as for $\{x_n : n \geq m\}$, so $a_m \leq a_{m+1}$ and $b_{m+1} \leq b_m$. Each $[a_m, b_m]$ is the smallest closed interval that contains the tail $\{x_n : n \geq m\}$. The diameter of the tail is the length of the interval, so $b_m - a_m$ tells us how much variation there is among its terms in the tail. We'll want the definition of convergent sequence to imply that $b_m - a_m \to 0$. When that's the case, our axioms guarantee that there is exactly one number L common to all the intervals $\{[a_m, b_m]\}_{m=1}^{\infty}$. Every term in the tail $\{x_n : n \geq m\}$ must be within $b_m - a_m$ of L, so in that case we can indeed think of every term in the tail $\{x_n : n \geq m\}$ as nearly the same as L whenever m is large enough.

Now let's give a formal definition that pins down these ideas. It's better if our definition omits any mention of the nested sequence $\{[a_m, b_m]\}_{m=1}^{\infty}$ because we may not have an effective method for calculating the supremum and infimum of any of the tails, let alone all of them.

DEFINITION 1.1:   A sequence $\{x_n\}_{n=1}^{\infty}$ of real numbers is a **convergent sequence** if there is a number L with the property that for each $\varepsilon > 0$ there is an integer M such that every x_n with $n \geq M$ satisfies $|x_n - L| < \varepsilon$. The number L is called the **limit** of the sequence $\{x_n\}_{n=1}^{\infty}$, and it is common to say that the sequence converges to its limit. We indicate this special relationship between the sequence and its limit symbolically as

$$\lim_{n \to \infty} x_n = L.$$

When we're primarily interested in displaying the value of the limit but not the nature of the limiting process involved, we may abbreviate it as $x_n \to L$.

In other words, a sequence converges to a limit if every open interval containing the limit also contains a tail of the sequence. A sequence can't possibly converge to more than one limit because we can certainly find disjoint open intervals around any two distinct numbers, and if one of those intervals contains a tail then the other one can't.

Now let's see that the definition we gave really does correspond to the ideas we discussed before giving it. We began our discussion by assuming that $\{x_n\}_{n=1}^{\infty}$ was a bounded sequence, but that doesn't seem to be part of the definition given. It is, however, a consequence, which is why it wasn't stated explicitly. For example, if we choose $\varepsilon = 1$ and then find M such that every x_n with $n \geq M$ satisfies $|x_n - L| < 1$, then $|x_n| \leq |L| + 1$ for all $n \geq M$, and we get a bound for the entire sequence:

$$|x_n| \leq \max\{|x_1|, |x_2|, \ldots, |x_{M-1}|, |L| + 1\} \quad \text{for all } n \in \mathbf{N}.$$

So every convergent sequence is necessarily bounded. Returning to the case of arbitrary ε, if every x_n with $n \geq M$ satisfies $|x_n - L| < \varepsilon$, then $L - \varepsilon$ and $L + \varepsilon$ are lower and upper bounds for the tail $\{x_n : n \geq M\}$. So we'll have

$$L - \varepsilon \leq \inf\{x_n : n \geq M\} = a_M \leq b_M = \sup\{x_n : n \geq M\} \leq L + \varepsilon.$$

Thus $b_M - a_M < 2\varepsilon$, and since ε is arbitrary our nested sequence of closed intervals will indeed satisfy $b_m - a_m \to 0$.

We can combine convergent sequences in a variety of ways to produce new ones. For example, if $\{x_n\}_{n=1}^{\infty}$ and $\{y_n\}_{n=1}^{\infty}$ are convergent sequences with $x_n \to L$ and $y_n \to L'$, then $x_n + y_n \to L + L'$ and $x_n y_n \to LL'$; those are consequences of the rules for approximate arithmetic. For example, if we're given $\varepsilon > 0$ and we want $|x_n y_n - LL'| < \varepsilon$, we know there is a $\delta > 0$ such that it's true whenever $|x_n - L| < \delta$ and $|y_n - L'| < \delta$. Invoking the assumed convergence of the two original sequences, there must be integers M and M' such that every x_n with $n \geq M$ satisfies $|x_n - L| < \delta$ and every y_n with $n \geq M'$ satisfies $|y_n - L'| < \delta$. Both conditions are satisfied for n greater than or equal to the larger of M and M', which proves $x_n y_n \to LL'$.

In addition to combining convergent sequences arithmetically to produce new convergent sequences, we may also use continuous functions for the same purpose. Let's look at how it works. Suppose that $x_n \to L$ and that f is a numerical function whose domain includes L as well as every term in the sequence. Then $\{f(x_n)\}_{n=1}^{\infty}$ is a new sequence of real numbers, and $f(x_n) \to f(L)$ if f is continuous at L. Again, this is quite easy to prove. When we're given $\varepsilon > 0$, we know there is a $\delta > 0$ such that every x in the domain of f with $|x - L| < \delta$ must satisfy $|f(x) - f(L)| < \varepsilon$. For $x_n \to L$, there is an integer N such that $x_n \in (L - \delta, L + \delta)$ for all $n \geq N$, and we've explicitly assumed that each x_n is in the domain of f.

Since we can work with convergent sequences so easily, most of the sequences we use will be convergent ones. But the ones we use are created with some care. A sequence constructed with less care may not converge, and then we call it **divergent**. There are many ways that a sequence of real numbers can fail to converge. There are certainly unbounded sequences, and since every convergent sequence is bounded we know that no unbounded sequence of real numbers can possibly converge. But bounded sequences of real numbers can also diverge; their terms can oscillate so that the diameters of the tails all stay above some positive value. The sequence defined by $x_n = (-1)^n$ is an especially simple example of a divergent bounded sequence. More generally, any time we can find an open interval (a, b) such that every tail of the sequence includes both terms below a and terms above b, the sequence can't converge; for $\varepsilon < \frac{1}{2}(b - a)$ the interval $(L - \varepsilon, L + \varepsilon)$ can never contain a tail of the sequence, no matter how we choose L.

EXERCISES

1. Let $\{x_n\}_{n=1}^{\infty}$ be a convergent sequence, and suppose that $x_n > 0$ for every n. Prove that $\lim_{n \to \infty} x_n \geq 0$.

2. Let $\{x_n\}_{n=1}^{\infty}$ and $\{y_n\}_{n=1}^{\infty}$ be convergent sequences, with $x_n \to a$ and $y_n \to b$. What must we assume to prove that $\{x_n/y_n\}_{n=1}^{\infty}$ is a sequence that converges to a/b?

3. Let $\{x_n\}_{n=1}^{\infty}$ be a convergent sequence, and let I be an open interval. Prove that if the sequence has infinitely many terms that are not in I, then $\lim_{n \to \infty} x_n \notin I$. Is there a similar result for closed intervals?

2 LIMITS AND REMOVING DISCONTINUITIES

Our condition for a discontinuity a of a numerical function f to be removable can be hard to check. We said that the discontinuity is removable if there is a function g whose domain includes both a and the domain of f, that is continuous at a, and that agrees with f everywhere except at a. That's fine when we can spot a way to define g by simply rewriting the formula for f, but how can we tell whether such a function g exists when there's no apparent formula for it? We really ought to have a condition stated in terms of f, not some unknown function g. And we ought to have a convenient way to refer to the value we assign to $f(a)$ to remove the discontinuity. That's the idea behind the following definition. Since our notion of continuity at a point depends on the point, the values of the function, and the domain of the function, our definition also depends on all these things.

DEFINITION 2.1:   Let f be a numerical function whose domain includes the set E, and let a be a point such that every interval $(a - \delta, a + \delta)$ with $\delta > 0$ contains points of E different from a. We say that the number L is the **limit of $f(x)$ as x approaches a through E** if for each $\varepsilon > 0$ there is a $\delta > 0$ such that every $x \in (a - \delta, a + \delta) \cap E$, with the possible exception of a itself, satisfies $|f(x) - L| < \varepsilon$. We indicate this condition symbolically by

$$\lim_{x \to_E a} f(x) = L.$$

The definition given in most calculus courses doesn't involve the set E, but there is a requirement that the domain of f include every $x \neq a$ in some open interval about a. However, the idea of restricting x to a smaller subset of values appears in the guise of one-sided limits, and that simple model helps us understand the definition given here. We'll return to the notion of one-sided limits at the end of this section.

It's important to recognize that our definition does specify the number L we call the limit. That is, if we have a point a, a set E, a function f whose domain includes E, and numbers L and L' that both satisfy the definition of the limit of $f(x)$ as x approaches a through E, then we must have $L = L'$. In the previous section we used the fact that every open interval about the limit of a sequence contains a tail of the sequence to prove that no sequence could have more than one limit, and we would like to do something similar here. In this case, sets of the form

$$\{f(x) : x \in (a - \delta, a + \delta) \cap E, \ x \neq a\}$$

play the same role as the tail of a sequence. Of course, the tail of an infinite sequence will always have at least one point in it because there will always be $n \geq m$. Similarly, every set of the form given above will have at least one point because of our assumption that every interval of the form $(a - \delta, a + \delta)$ contains points of E other than a.

It's clear that if L is the limit of $f(x)$ as x approaches a through E, and we define $g : E \cup \{a\} \to \mathbf{R}$ by $g(x) = f(x)$ for $x \neq a$ and $g(a) = L$, then g is continuous at a and agrees with f at all other points of $E \cup \{a\}$. Consequently, when E is the entire domain of f and $\lim_{x \to_E a} f(x)$ exists, then either f is already continuous at a and g is the same as f or f has a removable discontinuity at a and L is the value we use for $f(a)$ to remove it. Note that our definition makes no assumptions about whether a is a point in the domain of f, and it makes no direct mention of $f(a)$. That's as it should be; when f has a removable discontinuity at a, we think of the value $f(a)$ as either missing or incorrectly defined. However, when we know that f is continuous at a we can simply evaluate $f(a)$ to find the limit.

The bad news about our definition is that it doesn't tell us how to find L or even how to determine whether such a number can exist, and doing so can be very hard. While we can determine limits in many specific cases, in general the best we can do is to find alternative descriptions of the limit and the circumstances under which it exists. For example, when L is known to be the limit of $f(x)$ as x approaches a through E, then for sufficiently small $\delta > 0$ the set of values

$$\{f(x) : x \in E \text{ with } 0 < |x - a| < \delta\}$$

is a bounded, nonempty set of real numbers. For convenience, let's call c_δ its infimum and d_δ its supremum. Then the definition of limit guarantees us that for any $\varepsilon > 0$, there is a $\delta > 0$ such that $c_\delta \geq L - \varepsilon$ and $d_\delta \leq L + \varepsilon$. Conversely, the numbers c_δ and d_δ can be defined without knowing that the limit exists. It's not hard to show that the limit does exist if and only if there is exactly one number common to all the intervals $[c_\delta, d_\delta]$; that one number is the value of the limit. But that doesn't really tell us how to find L because we may not know exactly how to locate *any* of the intervals $[c_\delta, d_\delta]$, let alone all of them.

We can find statements of other conditions under which L must exist, but unless we assume additional hypotheses we won't have a systematic way to determine whether they are satisfied in general. Nonetheless, verifying such conditions for specific functions is a very real possibility. If we're clever enough, we can find a way to complete the task for just about every function that we can write down explicitly. But the methods we use invariably make use of special properties of the function at hand, which we must first recognize and then find a way to exploit. Recognizing an appropriate technique may be quite difficult, even if we understand the theory of limits perfectly. We're probably more likely to find a limit by recognizing a familiar way to remove a discontinuity than to remove a discontinuity by analyzing a limit.

Something that is easy to do with limits and functions in general is to recognize how the limits combine when we combine two or more functions. The definition of limit is designed to simplify this task and works beautifully with the rules for approximate arithmetic; most cases reduce to a single argument that gets used over and over again. If, for example, $\lim_{x \to_E a} f(x) = L_1$ and $\lim_{x \to_E a} g(x) = L_2$, then it's easy to see that $\lim_{x \to_E a} f(x) g(x) = L_1 L_2$. We can make $|f(x) g(x) - L_1 L_2|$ as small as necessary by simply making sure that $f(x)$ is close enough to L_1 and that $g(x)$ is close enough to L_2. Both of these conditions can be achieved by taking $x \neq a$ in $E \cap (x - \delta, x + \delta)$ with an appropriate $\delta > 0$. Once again, the key is to recognize that if it's possible to choose positive δ's to satisfy each of several conditions, then choosing the smallest of them allows us to satisfy all the conditions simultaneously.

Analyzing the limit is a little more complicated when a function of a function is involved. In considering $\lim_{x \to_E a} f(x)$, nothing is assumed about $f(a)$, which allows for some nasty surprises. To rule out such surprises, we need additional hypotheses beyond those we might expect, and there are several useful possibilities. The theorem below is a good example.

THEOREM 2.1: *Let A and B be sets of real numbers, with $f : A \to B$ and $g : B \to \mathbf{R}$. Suppose that $\lim_{x \to_A a} f(x) = b$ and that $\lim_{x \to_B b} g(x) = L$. Then $\lim_{x \to_A a} g(f(x)) = L$, provided either that b*

is a point of continuity for g or that there is a $\delta_0 > 0$ such that b is not one of the values of f on the set $\{x \in A : 0 < |x - a| < \delta_0\}$.

☐ *Proof:* We need to show that whenever $\varepsilon > 0$ is given, there is a $\delta > 0$ such that every $x \in (a - \delta, a + \delta) \cap A$ except possibly a itself must satisfy $|g(f(x)) - L| < \varepsilon$. Since we've assumed that $L = \lim_{x \to_B b}(x)$, we know that there is an $\eta > 0$ such that every $y \in (b - \eta, b + \eta) \cap B$ except possibly $y = b$ satisfies $|g(y) - L| < \varepsilon$. While we can pick $\delta > 0$ such that every $x \in A$ with $0 < |x - a| < \delta$ satisfies $f(x) \in (b - \eta, b + \eta) \cap B$, that may not be good enough. It is good enough when g is continuous at b because then $g(f(x)) = L$ when $f(x) = b$. Without continuity at b, we need to be able to choose $\delta > 0$ in such a way that no $x \in A$ with $0 < |x - a| < \delta$ satisfies $f(x) = b$. ∎

Discovering and proving theorems about limits of combinations of functions is basically a simple matter; all the difficulties in choosing a δ for a given ε are buried in the hypotheses assumed. Using the theorems is often so natural that as we break a complex function down into simpler components, we may not even be aware of the theorems involved. The skill comes in knowing when not to use the theorems; it's important to recognize when an expression with several parts must be dealt with as a unit.

In elementary calculus, we don't ordinarily deal with functions defined on arbitrary sets, and we're generally interested in the limit of a function at a point that is either an interior point or an endpoint of an interval in the domain of the function. Consequently, $\lim_{x \to_E a} f(x)$ is not ordinarily used. When E includes all $x \neq a$ in an open interval centered at a, the limit in question is called simply the **limit** of $f(x)$ as x approaches a, indicated symbolically as $\lim_{x \to a} f(x) = L$. We sometimes write this informally as $f(x) \to L$ as $x \to a$. When E is an interval (a, b), the limit is called the **right-hand limit** of $f(x)$ as x approaches a, indicated symbolically as $\lim_{x \to a+} f(x)$. When E is (c, a), the limit is called the **left-hand limit**, indicated as $\lim_{x \to a-} f(x)$. But all of these are simply special cases of a single concept, so a single theorem can deal with all three simultaneously, as we've done.

In fact, we sometimes investigate $\lim_{x \to a} f(x)$ by considering the left-hand and right-hand limits separately. This is especially useful when different computational formulas for $f(x)$ are involved, one used for $x < a$ and the other for $x > a$. Of course, $\lim_{x \to a} f(x)$ need not exist, but when it does it's a single number and has to be equal to both $\lim_{x \to a+} f(x)$ and $\lim_{x \to a-} f(x)$. While neither $\lim_{x \to a+} f(x)$ nor $\lim_{x \to a-} f(x)$ has to exist in general, when both exist but have different values we say that f

has a **simple jump discontinuity** at a. For example, $x/|x|$ has a simple jump discontinuity at the origin. Such discontinuities are easy to deal with as long as we stay on the same side of the jump since on either side they act like removable discontinuities.

EXERCISES

4. Let $f(x)$ be a nonnegative function, and suppose that a is a point for which $\lim_{x \to a} f(x)$ exists. Prove that $\lim_{x \to a} f(x) \geq 0$. If we make the stronger assumption $f(x) > 0$ for all x, can we conclude that $\lim_{x \to a} f(x) > 0$?

5. Prove that when $\lim_{x \to_E a} f(x) = L_1$ and $\lim_{x \to_E a} g(x) = L_2$, we must have $\lim_{x \to_E a} [f(x) + g(x)] = L_1 + L_2$.

6. Prove that if $p(x)$ is any polynomial, a is any real number, the function f defined by $[p(x) - p(a)] / (x - a)$ has a removable discontinuity at $x = a$. You'll need to make use of an important principle from algebra.

3   LIMITS INVOLVING ∞

As we study the graphs of given functions, we often check for the presence of horizontal asymptotes. There's an analytic version of this feature, too; the notion of the limit of a function at ∞ or $-\infty$. When there are horizontal asymptotes, the function values exhibit little change after we've moved far to the right or left. Just as we can use $\lim_{n \to \infty} x_n$ to approximate all the terms of the sequence $\{x_n\}_{n=1}^{\infty}$ with $n > N$, we can use $\lim_{x \to \infty} f(x)$ and $\lim_{x \to -\infty} f(x)$ to approximate the values of f for all x outside a closed interval $[a, b]$. That goal and our earlier work with limits pretty well tell us how these limits need to be defined, so we'll give their definition immediately.

DEFINITION 3.1:   We say that $\lim_{x \to \infty} f(x) = L$ provided that for each $\varepsilon > 0$, there is a $b \in \mathbf{R}$ such that every $x \in (b, \infty)$ is in the domain of f and satisfies $|f(x) - L| < \varepsilon$. Similarly, we say that $\lim_{x \to -\infty} f(x) = L$ provided that for each $\varepsilon > 0$, there is a real number a such that every $x \in (-\infty, a)$ is in the domain of f and satisfies $|f(x) - L| < \varepsilon$.

Of course, these limits don't necessarily exist, but when they're known to exist we can use that information in many ways. For example, if two or more functions are known to have limits at ∞ or at $-\infty$, we can immediately recognize the corresponding limits of many algebraic combinations of them. The proofs of such relationships are not significantly different from arguments given before; we just choose b large enough or a small enough to satisfy several conditions simultaneously.

To evaluate expressions of the form $\lim_{x \to \infty} [p(x)/q(x)]$ with p and q specific polynomials, many students look at the leading terms of the polynomials and ignore all the rest. Let's look at what's involved, and see when that simplified approach can be successful.

For $p(x)$ a polynomial of degree n, it is possible to write

$$p(x) = a_n x^n + a_{n-1} x^{n-1} + \cdots + a_1 x + a_0,$$

where the coefficients $\{a_k\}_{k=0}^n$ are constants with $a_n \neq 0$. We can always rewrite the formula for $p(x)$ as

$$\begin{aligned} p(x) &= x^n \left[a_n + a_{n-1} x^{-1} + \cdots + a_1 x^{1-n} + a_0 x^{-n} \right] \\ &= x^n P(1/x), \end{aligned}$$

where P is the polynomial defined by

$$P(y) = a_n + a_{n-1} y + \cdots + a_1 y^{n-1} + a_0 y^n.$$

We can't say the degree of P is n unless $a_0 \neq 0$, but we've assumed that

$$P(0) = a_n \neq 0.$$

Since all polynomials are continuous, P is continuous at 0. So when $\varepsilon > 0$ is given, there must be a $\delta > 0$ such that

$$\left| \frac{P(y)}{P(0)} - 1 \right| < \varepsilon \quad \text{for all } y \in (-\delta, \delta).$$

Consequently,

$$\left| \frac{p(x)}{a_n x^n} - 1 \right| = \left| \frac{P(1/x)}{a_n} - 1 \right| < \varepsilon \quad \text{for all } x \in (1/\delta, \infty),$$

which is the basis for approximating $p(x)$ by $a_n x^n$ as $x \to \infty$. Their ratio approaches 1, although their difference may be quite large.

The following theorem is a different sort of result that uses the existence of limits at $\pm\infty$ as a hypothesis and shows how we use those limits to reduce the study of a function on $(-\infty, \infty)$ to looking at what it does on a closed interval $[a, b]$.

THEOREM 3.1: *Let f be a continuous numerical function on $\mathbf{R}$, with $\lim_{x \to \infty} f(x) = R$ and $\lim_{x \to -\infty} f(x) = L$. Then f is a bounded func-*

tion, with $\sup \{f(x) : -\infty < x < \infty\}$ *and* $\inf \{f(x) : -\infty < x < \infty\}$ *either* $R,$ $L,$ *or values of* $f.$

□ *Proof:* Using $\varepsilon = 1$ in the definition of limit, we can choose a and b such that every x in $(-\infty, a)$ satisfies $|f(x) - L| < 1$, and every x in (b, ∞) satisfies $|f(x) - R| < 1$. Moreover, we can assume that $a < b$. Since f is continuous on $[a, b]$, we know there are numbers x_0 and x_1 in that closed interval such that

$$f(x_0) \le f(x) \le f(x_1) \quad \text{for all } x \in [a, b].$$

We also know that

$$L - 1 < f(x) < L + 1 \quad \text{for all } x \in (-\infty, a)$$

as well as

$$R - 1 < f(x) < R + 1 \quad \text{for all } x \in (b, \infty).$$

Consequently, the least of the three numbers $f(x_0)$, $L - 1$, and $R - 1$ is a lower bound for the values of f, and the greatest of the three numbers $f(x_1)$, $L + 1$, and $R + 1$ is an upper bound.

Having proved that f is a bounded function, we can define

$$A = \inf \{f(x) : x \in \mathbf{R}\} \quad \text{and} \quad B = \sup \{f(x) : x \in \mathbf{R}\},$$

and then $f(x) \in [A, B]$ for all x. The definition of limit won't allow either L or R to be below A or above B, so both must also be in $[A, B]$. Let's suppose that neither L nor R is B. In that case, we can find $\varepsilon > 0$ such that both $L + \varepsilon < B$ and $R + \varepsilon < B$, and then we can choose a' such that

$$f(x) < L + \varepsilon < B \quad \text{for every } x \in (-\infty, a')$$

as well as b' such that

$$f(x) < R + \varepsilon < B \quad \text{for every } x \in (b', \infty).$$

Since f is continuous on $[a', b']$, it has a greatest value on that interval, and that value must be B. Similarly, if neither L nor R equals A, then we can produce another interval $[a'', b'']$ on which the least value of f is A. ■

We conclude this section by noting that sometimes limits are said to be ∞ or $-\infty$; these are referred to as **infinite limits**. Some caution is needed to deal with infinite limits correctly. It is best to think of an infinite limit as only a convenient description of a property that a function without a limit may have. We can't use our rules for approximate arithmetic to work with algebraic combinations of functions with infinite limits since the symbols ∞ and $-\infty$ don't represent real numbers and are not subject to the usual laws of arithmetic. We can, however, use $\lim_{x \to \infty} f(x)$ or $\lim_{x \to -\infty} f(x)$ to evaluate the limit of $f(g(x))$ when g has an infinite limit; the definitions of infinite limits are designed to make this possible. Just as we say that $f(x) \to L$ if for every $\varepsilon > 0$ there are appropriate intervals in which every x satisfies $f(x) \in (L - \varepsilon, L + \varepsilon)$, we say $f(x) \to \infty$ if for every b there are appropriate intervals in which every x satisfies $f(x) \in (b, \infty)$. We make a similar modification to define $f(x) \to -\infty$. Of course, the exact definition depends on the sort of limit involved: two-sided limits, one-sided limits, limits at $-\infty$, or limits at ∞.

EXERCISES

7. For $f(x) = x\,|x| / (x^2 + 1)$, find the supremum and infimum of the set of values of f. Are they values of f?

8. Suppose that f is a numerical function defined on $(0, \infty)$. While the limit of the sequence $\{f(n)\}_{n=1}^{\infty}$ and $\lim_{x \to \infty} f(x)$ often refer to the same number, they are actually different concepts. What is the relationship between them?

9. Prove that if $\lim_{x \to \infty} f(x) = L_1$ and $\lim_{x \to \infty} g(x) = L_2$, then

$$\lim_{x \to \infty} [f(x) + g(x)] = L_1 + L_2.$$

10. Prove that if $\lim_{x \to \infty} f(x) = L$, then $\lim_{x \to 0+} f(1/x) = L$ as well.

11. Formulate a precise definition of $\lim_{x \to \infty} f(x) = \infty$.

12. Prove that if $\lim_{x \to \infty} f(x) = L$ and $\lim_{x \to \infty} g(x) = \infty$, then

$$\lim_{x \to \infty} [f(x) + g(x)] = \infty.$$

13. Prove that if $\lim_{x \to \infty} |f(x)| = \infty$, then $\lim_{x \to \infty} 1/f(x) = 0$.

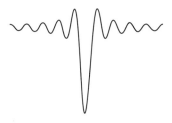

IV

THE DERIVATIVE

Since calculus developed largely to meet the needs of science, there are many parallels between scientific methodology and calculus concepts. For the most part, science proceeds by studying change, and the study of changes is the basis for differential calculus.

I DIFFERENTIABILITY

The first step in any serious study of change is to find a way to measure it. That's especially easy to do with numerical functions. The quantity $f(x) - f(a)$ represents a change in the value of the function f, and comparing this change to $x - a$ gives us a convenient scale for measuring it. Many elementary functions exhibit a common pattern:

$$f(x) - f(a) = (x - a) g(x), \quad \text{with } g \text{ continuous at } a. \tag{4.1}$$

For example, if f is defined by any polynomial, then $f(x) - f(a)$ is a polynomial that vanishes when $x = a$, and so by the factor theorem we know that $(x - a)$ is a factor of $f(x) - f(a)$. The complementary factor $g(x)$ in (4.1) is also a polynomial, and we know that polynomials define continuous functions on the entire real line. When f isn't a polynomial, representations of the form (4.1) may still be possible, but the continuity of g at a is harder to establish.

It is the continuity of g at a that makes equation (4.1) useful because it lets us approximate $g(x)$ by $g(a)$ when x is near to a. To use this approximation more effectively, we rewrite equation (4.1) as

$$f(x) = f(a) + (x - a)g(a) + (x - a)[g(x) - g(a)].$$

The sum $f(a) + (x - a)g(a)$ forms the **linearization** of $f(x)$ near $x = a$, and the term $(x - a)[g(x) - g(a)]$ is thought of as a correction to the linearization. For g continuous at a, the correction term is small in comparison to $|x - a|$ when x is near a. More precisely, for any given $\varepsilon > 0$ there is an interval $(a - \delta, a + \delta)$ on which the correction is never larger than $\varepsilon |x - a|$. The linearization of f is quite easy to understand and to work with, so when we can ignore the effect of the correction term we can eliminate all the difficulties that complicated functions can introduce. This is the basic idea behind **differentiability**, the concept defined below.

DEFINITION 1.1: For f a numerical function, we say the function f is **differentiable** at the point a if the domain of f includes an open interval I about a, and there is a function $g : I \to \mathbf{R}$ that is continuous at a and satisfies

$$f(x) = f(a) + (x - a)g(x) \quad \text{for all } x \in I.$$

When f is differentiable at a, the function g in our definition must be defined at each $x \neq a$ in I by the formula

$$g(x) = \frac{f(x) - f(a)}{x - a}. \tag{4.2}$$

We may note that whether or not f is differentiable at a, the function g defined by (4.2) is continuous at all the other points of continuity of f. Of course, the quotient in (4.2) has a discontinuity at a since we're assuming it is defined at all the other points in I. Then f is differentiable at a precisely when we can remove the discontinuity at a. In that case,

$$g(a) = \lim_{x \to a} \frac{f(x) - f(a)}{x - a}. \tag{4.3}$$

So another way to define differentiability is to say that f is differentiable at a provided the limit in (4.3) exists. That's the way differentiability is usually defined in introductory calculus books. When the limit exists, it's indicated by $f'(a)$ and called the **value of the derivative of f at a**.

While it's natural to emphasize this limit when the goal is finding derivatives, we've taken a somewhat different point of view to emphasize a different aspect of the theory. Our goal is to develop properties of functions known to be differentiable, so we're adopting the definition in a form that simplifies their discovery and verification. For example, when f is differentiable at a, the way we've expressed $f(x)$ in terms of continuous functions makes it obvious that f is continuous at a.

Our definition of differentiability makes it easy to appreciate the reason that derivatives are used to search for extreme values of functions. When f is differentiable at a with $f'(a) \neq 0$, the continuity of g at a in the representation

$$f(x) - f(a) = (x - a) g(x)$$

requires $g(x)$ to have the same sign as $f'(a)$ at all points in some open interval $(a - \delta, a + \delta)$, so in that interval the values of $f(x) - f(a)$ are positive on one side of a and negative on the other. So when $f'(a)$ exists and is nonzero, we see that $f(a)$ is neither the largest nor the smallest value of $f(x)$. Points in the domain of f at which f' is either undefined or zero are called **critical points**; when we're searching for extreme values in an open interval we can confine our search to critical points.

While our concept of differentiability really is the same as that given in most introductory calculus books, a meaningful extension of the concept is possible and is sometimes used. It isn't entirely necessary to insist that the domain of f include an open interval I about a; when the domain of f is a closed interval some authors allow a to be an endpoint. We've chosen not to allow that. Our point of view is that a function differentiable at a should closely resemble a linear function near $x = a$, and linear functions are defined at all the points near a, not just some of them.

As we mentioned above, we generally indicate

$$\lim_{x \to a} \frac{f(x) - f(a)}{x - a} = f'(a).$$

We think of $f'(a)$ as the value of a new function f' derived from f, called the **derivative** of f. The function f' is not the same as the function g in our definition of differentiability at the point a; note f' and g have the same value at a but not necessarily at any other points. If $f : E \to \mathbf{R}$, then the domain of f' is the set of points in E at which f is differentiable. We call f a **differentiable function** if it is differentiable at all points in its domain, and in that case f and f' have the same domain. The operation we perform on f to create the function f' is called **differentiation**; it is an operation that we perform on functions, not on numbers. Consequently, two functions that have the same value at $x = a$ won't necessarily have equal derivatives at $x = a$. It is true that equal functions have equal derivatives, but equal functions must have equal values at all points, not just at a.

When f is differentiable at all points of an interval, it's natural to expect f' to be continuous there since its values are found by removing discontinuities. That happens to be true for many of the functions we work with, but it's false in general. The discontinuity we remove to find $f'(a)$ isn't a discontinuity of $f'(x)$, it's a discontinuity of another function, and f' may well have a nonremovable discontinuity at points in its domain. Nonetheless, when f' is defined on an interval it does indeed have some of the properties we associate with continuous functions. We'll investigate this point in the third section of this chapter.

It's common to interpret differentiability geometrically in terms of the existence of a line tangent to a curve. Without worrying too much about when a set of points forms a curve, it is now agreed that a line $\mathcal{L}$ through a point P on a curve $\mathcal{C}$ is a tangent line provided that whenever a moving point Q on $\mathcal{C}$ approaches P, the angle between $\mathcal{L}$ and the line through P and Q approaches zero. The directions of nonvertical lines are determined by their slopes, so we say that the nonvertical line $\mathcal{L}$ through P is the tangent line at P if the slope of the line through P and Q approaches the slope of $\mathcal{L}$ as the moving point Q approaches P along $\mathcal{C}$. If our curve is the graph of f and P is the point $(a, f(a))$, we may take Q to be $(x, f(x))$ with $x \neq a$. Then the slope of the line through P and Q is exactly

$$\frac{f(x) - f(a)}{x - a},$$

and when it has a limit f is differentiable at a.

There are, in fact, other ways that we can specify a curve without giving y explicitly as a function of x, and it may still be possible to find the slope of a tangent line by taking a limit of slopes. If we think of (x, y) as the fixed point P on $\mathcal{C}$ and $(x + \Delta x, y + \Delta y)$ as the moving point Q, then the slope of the line through P and Q is $\Delta y / \Delta x$, with Δx and Δy both approaching zero as Q approaches P. The slope of the tangent line at P is then

$$\frac{dy}{dx} = \lim_{\Delta x \to 0} \frac{\Delta y}{\Delta x},$$

provided this limit exists.

The symbol $\frac{dy}{dx}$ for the slope of a tangent line was introduced by Gottfried Leibniz, who discovered many of the basic principles of calculus at about the same time as Newton. When $y = f(x)$ with f a differentiable function, $\frac{dy}{dx}$ and $f'(x)$ have exactly the same meaning and are used almost interchangeably. However, there's no convenient way to indicate the coordinates of the point of tangency in Leibniz's notation. In some situations, that's a definite disadvantage. But the $f'(x)$ notation can be used only when some function named f has been defined, and sometimes we find the derivative before we get around to introducing f. In such cases, Leibniz's notation is more convenient.

Our notation $f'(x)$ for the derivative wasn't Newton's choice either; in fact, he didn't usually think of y as a function of x. Instead, he preferred to think of (x, y) as the coordinates of a moving point, with both x and y functions of time. Assuming that $y = f(t)$ and $x = g(t)$, we have

$$\Delta y = f(t + \Delta t) - f(t) \quad \text{and} \quad \Delta x = g(t + \Delta t) - g(t).$$

Newton assumed that x and y were differentiable functions of t, with derivatives

$$\dot{x} = f'(t) \quad \text{and} \quad \dot{y} = g'(t).$$

As long as $\dot{x} \neq 0$, the slope of the tangent line is

$$\lim_{\Delta t \to 0} \frac{\Delta y}{\Delta x} = \lim_{\Delta t \to 0} \frac{\Delta y / \Delta t}{\Delta x / \Delta t} = \frac{\dot{y}}{\dot{x}},$$

which was Newton's notation for the slope of the tangent line.

EXERCISES

1. Use the definition of differentiability to show that if f is differentiable at the point a, then so is f^2.

2. By considering the cases $a > 0$, $a = 0$, and $a < 0$ separately, show that $f(x) = x|x|$ defines a function that is differentiable at each $a \in \mathbf{R}$.

3. Use the definition of differentiability to show that

$$f(x) = \begin{cases} x + x^2 \sin(1/x), & x \neq 0 \\ 0, & x = 0 \end{cases}$$

    defines a function that is differentiable at 0, with $f'(0) = 1$.
    Note: Everywhere except at the origin, the derivative of f is given by

$$f'(x) = 1 + 2x \sin(1/x) - \cos(1/x).$$

The discontinuity of f' at 0 is not removable, even though $f'(0)$ exists.

4. Show that if (x, y) and $(x + \Delta x, y + \Delta y)$ are points on the circle $x^2 + y^2 = r^2$ with y, Δx, and $y + \Delta y$ nonzero, then

$$\frac{\Delta y}{\Delta x} = -\frac{2x + \Delta x}{2y + \Delta y} \quad \text{and} \quad \frac{dy}{dx} = -\frac{x}{y}.$$

2   COMBINING DIFFERENTIABLE FUNCTIONS

We've seen that when we combine continuous functions in a number of ways, the new functions we produce are also continuous. We'd like to be able to recognize that the same sorts of combinations of differentiable functions are differentiable. We'll see how that works by first deriving rules for the derivative of basic arithmetic combinations of differentiable functions. Then we'll give a formal statement and rigorous proof of the chain rule. The formulas we derive in this section should be familiar from basic calculus; what may be unfamiliar is the way the formulas follow naturally from the definition of differentiable functions and derivatives.

Suppose we have two functions f_1 and f_2 that are known to be differentiable at the same point a, and that f is a new function defined in terms of f_1 and f_2. First we note that there must be an open interval I_1 containing a and a function g_1 that is continuous at a and satisfies

$$f_1(x) = f_1(a) + (x - a)g_1(x) \quad \text{for all } x \in I_1. \tag{4.4}$$

Also, there is an open interval I_2 containing a and a function g_2 that is continuous at a and satisfies

$$f_2(x) = f_2(a) + (x - a)g_2(x) \quad \text{for all } x \in I_2. \tag{4.5}$$

Then $I = I_1 \cap I_2$ defines an open interval containing a on which both equations are true. Our strategy is to use equations (4.4) and (4.5) to express

$$f(x) = f(a) + (x - a)g(x) \quad \text{for all } x \in I$$

with $g(x)$ defined in terms of $f_1(a)$, $f_2(a)$, $g_1(x)$, and $g_2(x)$. Whenever we can recognize g as continuous at a, we'll know that f is differentiable at a with $f'(a) = g(a)$.

First we consider linear combinations of f_1 and f_2. For this case we define $f : I \to \mathbf{R}$ by

$$f(x) = c_1 f_1(x) + c_2 f_2(x)$$

with c_1 and c_2 unspecified constants. For all $x \in I$, we may write

$$
\begin{aligned}
f(x) &= c_1 \left[f_1(a) + (x-a) g_1(x) \right] + c_2 \left[f_2(a) + (x-a) g_2(x) \right] \\
&= c_1 f_1(a) + c_2 f_2(a) + (x-a) \left[c_1 g_1(x) + c_2 g_2(x) \right] \\
&= f(a) + (x-a) g(x),
\end{aligned}
$$

with g defined by

$$
g(x) = c_1 g_1(x) + c_2 g_2(x).
$$

Since g is continuous at a whenever g_1 and g_2 are, we recognize that f is differentiable at a. Moreover, since

$$
g(a) = c_1 g_1(a) + c_2 g_2(a),
$$

we've proved that

$$
(c_1 f_1 + c_2 f_2)'(a) = c_1 f_1'(a) + c_2 f_2'(a).
$$

Next we consider the product of f_1 and f_2. This time we define the numerical function f on I by

$$
f(x) = f_1(x) f_2(x).
$$

Then for all $x \in I$ we may write

$$
\begin{aligned}
f(x) &= \left[f_1(a) + (x-a) g_1(x) \right] \left[f_2(a) + (x-a) g_2(x) \right] \\
&= f_1(a) f_2(a) + (x-a) f_1(a) g_2(x) \\
&\quad + (x-a) g_1(x) f_2(a) + (x-a)^2 g_1(x) g_2(x) \\
&= f(a) + (x-a) g(x),
\end{aligned}
$$

with g the numerical function on I defined by

$$
g(x) = f_1(a) g_2(x) + g_1(x) f_2(a) + (x-a) g_1(x) g_2(x).
$$

Again we recognize g as continuous at a, proving that f is differentiable at a. Since

$$
g(a) = f_1(a) g_2(a) + g_1(a) f_2(a),
$$

we've established the **product rule**

$$
(f_1 f_2)'(a) = f_1(a) f_2'(a) + f_1'(a) f_2(a).
$$

Now we consider dividing f_1 by f_2. Since we can't divide by zero, we should assume that $f_2(a) \neq 0$. Since f_2 is continuous at a, we can assume that our open interval I_2 was chosen in such a way that $f_2(x)$ is never 0 on I_2, and then we define a numerical function f on I by

$$f(x) = \frac{f_1(x)}{f_2(x)} = \frac{f_1(a) + (x-a)g_1(x)}{f_2(a) + (x-a)g_2(x)}.$$

Regrouping this last expression in the form $f(a) + (x-a)g(x)$ is a little trickier here. We get

$$g(x) = \frac{f(x) - f(a)}{x - a}$$

$$= \frac{1}{(x-a)}\left[\frac{f_1(a) + (x-a)g_1(x)}{f_2(a) + (x-a)g_2(x)} - \frac{f_1(a)}{f_2(a)}\right]$$

$$= \frac{[f_2(a)g_1(x) - f_1(a)g_2(x)]}{[f_2(a) + (x-a)g_2(x)]f_2(a)}$$

because the terms involving the product of $f_1(a)$ and $f_2(a)$ cancel out when we subtract. Since $f_2(a) \neq 0$, we again see that g is continuous at a, proving that f is differentiable at a. Since

$$g(a) = \frac{f_2(a)g_1(a) - f_1(a)g_2(a)}{[f_2(a)]^2},$$

we've established the **quotient rule**

$$\left(\frac{f_1}{f_2}\right)'(a) = \frac{f_2(a)f_1'(a) - f_1(a)f_2'(a)}{[f_2(a)]^2}.$$

That takes care of the rules for differentiating linear combinations, products, or quotients of differentiable functions. We still need to treat the composition of differentiable functions; that's the subject of the **chain rule**. We'll need a new set of assumptions here, so we'll state them formally.

THEOREM 2.1:   *Suppose that f is a function that is differentiable at the point a, and that F is a function that is differentiable at the point $f(a)$. Then there is an open interval I containing the point a on which the composition*

$$H(x) = F(f(x))$$

defines a function H that is differentiable at the point a. The value of its derivative there is

$$H'(a) = F'(f(a))f'(a).$$

☐ *Proof:* Since f is differentiable at a, we know there is an open interval I_0 about a and a function $g : I_0 \to \mathbf{R}$ that is continuous at a, and that satisfies

$$f(x) = f(a) + (x - a) g(x) \quad \text{for all } x \in I_0. \tag{4.6}$$

Also, since F is differentiable at $f(a)$, we know there is an open interval I_1 about $f(a)$ and a function $G : I_1 \to \mathbf{R}$ that is continuous at $f(a)$, and that satisfies

$$F(y) = F(f(a)) + [y - f(a)] G(y) \quad \text{for all } y \in I_1. \tag{4.7}$$

Our plan is to substitute $y = f(x)$ in (4.7) with $f(x)$ given by (4.6), but first we need to choose the interval I in a way that makes this substitution valid for all $x \in I$.

Since I_1 is an open interval about $f(a)$, there must be an $\varepsilon > 0$ such that $(f(a) - \varepsilon, f(a) + \varepsilon) \subset I_1$, and since f is continuous at a, there must be a $\delta > 0$ such that

$$|f(x) - f(a)| < \varepsilon \quad \text{for all } x \in I_0 \cap (a - \delta, a + \delta).$$

Then for each $x \in I = I_0 \cap (a - \delta, a + \delta)$, we get $f(x) \in I_1$ and therefore

$$H(x) = F(f(x)) = F(f(a)) + [f(x) - f(a)] G(f(x))$$

by (4.7). Since $F(f(a)) = H(a)$ and $f(x)$ satisfies (4.6), we see that

$$H(x) = H(a) + (x - a) g(x) G(f(x)) \quad \text{for all } x \in I.$$

Since $g(x) G(f(x))$ is continuous at a, we see that H is differentiable at a with

$$H'(a) = g(a) G(f(a)) = f'(a) F'(f(a)).$$

That completes our proof of the chain rule. ∎

We should note that the chain rule gives sufficient conditions for a function of a function to be differentiable; the conditions are by no means necessary. For example, $f(x) = \sqrt[3]{x}$ is defined and continuous for all x, but it is not differentiable at $x = 0$. Clearly,

$$\frac{f(x) - f(0)}{x - 0} = \frac{x^{1/3}}{x} = x^{-2/3}$$

and $x^{-2/3}$ has an infinite discontinuity at 0, not a removable one. Yet for $F(x) = x^3$,

$$H(x) = F(f(x)) = \left(\sqrt[3]{x}\right)^3 = x \quad \text{for all } x,$$

and so H is clearly differentiable at 0 with $H'(0) = 1$.

Sometimes it is easier to work with an identity involving the values of a function than with an explicit formula for the function itself. In such cases, we may be able to express $f(x)$ in the form $f(a) + (x - a)g(x)$ by defining g in terms of f rather than giving a complete formula for it. In these cases, the continuity of f becomes an important part of showing that g is continuous. We will see systematic ways of exploiting this idea in Sections 4.6 and 4.7, and it is the key to Exercise 8 below.

EXERCISES

5. According to the chain rule, if $F(x) = f(x^2)$ and f is differentiable at a^2, then F is differentiable at a with $F'(a) = 2af'(a^2)$. Use the definition to prove this directly.

6. If the numerical function f is differentiable at a and $F(x) = 1/f(x)$, then $F'(a) = -f'(a)/f(a)^2$ provided that $f(a) \neq 0$. Show that this follows from the quotient rule and also from the chain rule.

7. Suppose that f and F are functions such that $F(f(x)) = x$ for all x in an open interval I. Show that if a is any point in I such that f is differentiable at a and F is differentiable at $f(a)$, then neither $F'(f(a))$ nor $f'(a)$ is 0.

8. Suppose that $F(f(x))$ defines a function that is differentiable at a. Prove that if f is continuous at a and $f(a)$ is not a critical point for F, then f is differentiable at a.

3   MEAN VALUES

In formal mathematics, **mean** is often used instead of the more familiar term *average*, probably to alert the reader that a precisely defined mathematical concept is intended rather than the loose idea conveyed by phrases such as "average person" or "average day." When we talk about the mean value of a function, it's important to realize that we're referring to the mean of a set of values of f, not necessarily to a value of the function. We've seen a similar distinction in our study of the intermediate value theorem, where the goal was to prove that each number intermediate between two values of a continuous function is also a value of the function.

When f is defined on $[a, b]$, the quantity

$$m = \frac{f(b) - f(a)}{b - a}$$

is sometimes thought of as representing the mean value of the derivative of f over $[a, b]$. That's because the net change of f over $[a, b]$ is the same as that for mx, and the derivative of mx is always equal to m. This is analogous to the case when the interval $[a, b]$ represents a time interval and the values of f represent position. In that case, values of $f'(t)$ are called instantaneous velocities and m is called the average velocity over the time interval.

Clearly some hypotheses must be satisfied before we can assert that the mean value of a function is actually one of its values. It's reasonable to expect that continuity is involved somehow since we needed that assumption to prove the intermediate value theorem. The theorem below, universally known as the **mean value theorem**, gives conditions under which the mean value of a derivative is actually a value of the derivative. Note that the hypotheses are surprisingly weak; continuity of the derivative need not be assumed. However, the function itself must be continuous.

THEOREM 3.1: *Let f be a numerical function that is continuous on $[a, b]$ and differentiable at every point of (a, b). Then there is a point $\xi \in (a, b)$ such that*

$$f'(\xi) = \frac{f(b) - f(a)}{b - a}.$$

The mean value theorem is both deep and powerful. Many newer calculus books are attempting to leave it out. There are indeed more intuitive explanations available for many of its traditional applications; none, however, can rival the mean value theorem for simplicity and power. We'll use it in just about every remaining section of this book.

□ *Proof:* We'll follow the route taken by most calculus texts that include the mean value theorem, and justify it by appealing to an intermediate result, generally called **Rolle's theorem.** Rolle's theorem states that if $a < b$ and f is continuous on $[a, b]$ with $f(a) = f(b)$, then f has a critical point in (a, b). To prove it, we observe that for f continuous on $[a, b]$, the extreme value theorem tell us that f has both a least and greatest value over that interval. We've assumed that $f(a) = f(b)$, so f must have at least one of its extreme values at an interior point of $[a, b]$, and such an interior point must be a critical point.

It's a simple step from Rolle's theorem to the mean value theorem. If f is continuous on $[a, b]$ and we define $F : [a, b] \to \mathbf{R}$ by

$$F(x) = (b - a) f(x) - (x - a) [f(b) - f(a)],$$

then clearly F is continuous and

$$F(b) = (b - a) f(b) - (b - a) [f(b) - f(a)] = (b - a) f(a) = F(a).$$

So Rolle's theorem guarantees that F has a critical point $\xi \in (a, b)$. Clearly, F is differentiable everywhere that f is, with

$$F'(x) = (b - a) f'(x) - [f(b) - f(a)].$$

Since f has been assumed to be differentiable on (a, b), F is differentiable at its critical point ξ, and so

$$0 = F'(\xi) = (b - a) f'(\xi) - [f(b) - f(a)].$$

Solving for $f'(\xi)$ completes the proof.   ■

The simplicity of these arguments raises an obvious question: what's so deep about the mean value theorem? Certainly there's nothing at all remarkable about the step from Rolle's theorem to the mean value theorem. The difficulty is in proving Rolle's theorem, and the only place where that argument is not routine is in establishing the existence of a point ξ in (a, b) where f has an extreme value. That's what makes the mean value theorem a deep result. When we proved that every function continuous on a closed interval had both a greatest and a least value there, we were unable to give a general method for locating the points where they occurred. The statement of the mean value theorem gives no hint as to how the point ξ can be located, nor does the proof we gave provide any clues to finding it.

There are proofs of the mean value theorem that make no use of extreme values, and the problem of locating ξ is simpler than locating an extreme value. But we won't pursue the search for ξ, since our ability to locate it doesn't add to the usefulness of the mean value theorem. In fact, when we do find ξ we don't need the mean value theorem; a theorem is never needed to demonstrate the existence of something already at hand. The point of the theorem is that we can use the range of values of f' over (a, b) to estimate $f(b) - f(a)$ as a multiple of $b - a$. It may not be necessary to know $f(a)$ and $f(b)$ or even a and b to do so. The simplest case is when f' is 0 everywhere. In that case, we conclude that f must be constant. When f' keeps the same sign throughout (a, b), the mean value theorem tells us that $f(b) - f(a)$ shares that sign.

One remarkable consequence of the mean value theorem is that when f is differentiable on an interval I, the set of values $\{f'(x) : x \in I\}$ is also an interval, even when f' has discontinuities in I. To prove this, suppose that $[a, b] \subset I$, and that r is any number strictly between $f'(a)$ and $f'(b)$. Since f must be continuous on I, for $0 < h < b - a$ the formula

$$s_h(x) = \frac{f(x+h) - f(x)}{h}$$

defines a continuous function s_h on $[a, b - h]$. Since

$$\lim_{h \to 0} s_h(a) = f'(a) \quad \text{and} \quad \lim_{h \to 0} s_h(b - h) = f'(b),$$

if we take h small enough then r will be between $s_h(a)$ and $s_h(b - h)$ as well as between $f'(a)$ and $f'(b)$. For such an h, the intermediate value theorem guarantees there is a $c \in (a, b - h)$ with

$$r = s_h(c) = \frac{f(c+h) - f(c)}{h}.$$

Then the mean value theorem guarantees there is a $\xi \in (c, c + h) \subset (a, b)$ with

$$f'(\xi) = \frac{f(c+h) - f(c)}{h} = r.$$

The next theorem illustrates a more typical use of the mean value theorem. It shows that derivatives must have another property that we generally associate with continuous functions, even though derivatives are allowed to have discontinuities.

THEOREM 3.2: *Let a be a point in a given open interval I, and let f be a continuous function on I. If f is known to be differentiable at all points of I except possibly at a and $\lim_{x \to a} f'(x)$ exists, then f must also be differentiable at a with*

$$f'(a) = \lim_{x \to a} f'(x).$$

□ *Proof:* We define $g(x) = [f(x) - f(a)] / (x - a)$ for $x \in I$ but $x \neq a$, and then note that the conclusion of the theorem is simply that

$$\lim_{x \to a} g(x) = \lim_{x \to a} f'(x).$$

Calling L the limit of f', we need to show that when $\varepsilon > 0$ is given, there must be a $\delta > 0$ such that every $x \in (a - \delta, a + \delta)$ except possibly a itself satisfies $|g(x) - L| < \varepsilon$. We know there is a δ such that $|f'(x) - L| < \varepsilon$ for all $x \neq a$ in $(a - \delta, a + \delta)$, but f' and g are different functions. Fortunately, their values are strongly related. For $x \in I$ but $x \neq a$, we can apply the mean value theorem to f on $[x, a]$ when $x < a$ and on $[a, x]$ when $a < x$; in either case, we see there must be a point ξ between x and a with

$$f'(\xi) = \frac{f(a) - f(x)}{a - x} = \frac{f(x) - f(a)}{x - a} = g(x).$$

So if all the values of f' on $(a - \delta, a + \delta)$ are between $L - \varepsilon$ and $L + \varepsilon$, the same must be true of the values of g.   ■

We should note that the mean value theorem only works one direction. For a given ξ, the value $f'(\xi)$ may not be the slope of any line segment joining two points on the graph. The simplest example involves $f(x) = x^3$; obviously $f'(0) = 0$. However, for $a \neq b$ we see

$$\frac{f(b) - f(a)}{b - a} = b^2 + ab + a^2 = \frac{1}{2}\left[a^2 + b^2 + (a + b)^2\right],$$

and that must be positive. Theorem 3.2 only works one way for the same reason, so the formula for $f'(x)$ when $x \neq a$ isn't always useful for finding $f'(a)$.

Here's another theorem that can be very useful. It's sometimes called Cauchy's mean value theorem, in honor of the French mathematician Augustin Cauchy (1789–1857). Cauchy's name is associated with many fundamental concepts in the mathematics that developed from calculus, including the notion of limits. One of the uses of this theorem is to prove L'Hôpital's rule, a method for using derivatives to find limits instead of using limits to find derivatives. We'll develop L'Hôpital's rule in Chapter 9.

THEOREM 3.3: *Suppose that f and g are continuous on $[a, b]$ and differentiable on (a, b), with g' never vanishing there. Then there is a point $\xi \in (a, b)$ such that*

$$\frac{f(b) - f(a)}{g(b) - g(a)} = \frac{f'(\xi)}{g'(\xi)}.$$

□ *Proof:* Before we begin, we note that the mean value theorem implies that $g(b) - g(a) \neq 0$ since $g'(x) \neq 0$ for $a < x < b$. Thus the conclusion of the theorem does not involve division by zero.

Cauchy's mean value theorem is also a simple consequence of Rolle's theorem. This time we define the auxiliary function $F : [a, b] \to \mathbf{R}$ by the formula

$$F(x) = [f(b) - f(a)] g(x) - [g(b) - g(a)] f(x).$$

It's obviously continuous on $[a, b]$ and differentiable on (a, b), so any critical point for F in (a, b) must be a point where F' vanishes. When we evaluate either $F(a)$ or $F(b)$, some terms cancel and we find

$$F(a) = f(b) g(a) - f(a) g(b) = F(b).$$

Hence Rolle's theorem guarantees that there is a point $\xi \in (a, b)$ with

$$0 = F'(\xi) = [f(b) - f(a)] g'(\xi) - [g(b) - g(a)] f'(\xi).$$

Since neither $g(b) - g(a)$ nor $g'(\xi)$ is zero, this equation is equivalent to the statement of the theorem.   ∎

EXERCISES

9. Use the mean value theorem to prove that for $a < b$,

$$\left| \sqrt{1 + b^2} - \sqrt{1 + a^2} \right| < b - a.$$

10. Use the mean value theorem to prove that if $0 \le a < b$ then

$$3a^2 (b - a) < b^3 - a^3 < 3b^2 (b - a).$$

Why is the restriction $0 \le a$ necessary?

11. Suppose that f and g are positive, differentiable functions on an interval I, with

$$\frac{f'(x)}{f(x)} < \frac{g'(x)}{g(x)} \quad \text{for all } x \in I.$$

Prove that $F(x) = f(x) / g(x)$ defines a decreasing function on I.

12. Under the hypotheses of Theorem 3.3, show that

$$\lim_{x \to a+} \frac{f(x) - f(a)}{g(x) - g(a)} = \lim_{x \to a+} \frac{f'(x)}{g'(x)}$$

when the second limit exists. This is one of the arguments used to establish L'Hôpital's rule.

13. By considering the intervals between successive roots, prove that if $p(x)$ is a nonzero polynomial that vanishes at n different points, then there must be at least $n - 1$ points where $p'(x)$ vanishes. Then use induction to prove that no polynomial of degree n can have more than n roots.

4   SECOND DERIVATIVES AND APPROXIMATIONS

When the numerical function f is differentiable at the point a, we know that its domain includes an open interval I about a, and that the function $g : I \to \mathbf{R}$ defined by the formula

$$g(x) = \begin{cases} \dfrac{f(x) - f(a)}{x - a}, & x \neq a \\ f'(a), & x = a \end{cases}$$

is continuous at a. If g is itself differentiable at a, then there must be another function $h : I \to \mathbf{R}$ that is continuous at a and satisfies

$$g(x) = f'(a) + (x - a) h(x) \quad \text{for all } x \in I.$$

For $x \in I$, we can combine the formulas for $f(x)$ and $g(x)$:

$$f(x) = f(a) + (x - a) g(x)$$
$$= f(a) + (x - a) f'(a) + (x - a)^2 h(x).$$

Rewriting

$$h(x) = h(a) + [h(x) - h(a)],$$

we obtain

$$f(x) = f(a) + (x - a) f'(a) + (x - a)^2 h(a)$$
$$+ (x - a)^2 [h(x) - h(a)].$$

The first three terms in this last expression define a simple quadratic polynomial, and the last term represents the error when we use this polynomial to approximate $f(x)$. Since h is continuous at a, the last term becomes smaller than any nonzero constant multiple of $(x - a)^2$ as x approaches a. So we generally expect this quadratic approximation to be much more accurate than the linearization of $f(x)$ near $x = a$. Quadratic functions are fairly easy to work with, so such an approximation is potentially quite useful. To make quadratic approximations easy to use, we need a simple way to verify that g is differentiable at a and to calculate $g'(a)$.

Since $g(a) = f'(a)$, for all $x \in I$ except a we must have

$$h(x) = \frac{f(x) - f(a) - (x - a) f'(a)}{(x - a)^2},$$

an expression that clearly has a discontinuity at $x = a$. We can remove the discontinuity when f satisfies some very simple conditions. If we assume that f is differentiable not just at a but everywhere between a and x, and that it is continuous at x as well as at a and at all the points between a and x, then we can simplify the formula for $h(x)$ with the aid of Cauchy's mean value theorem. We think of h as given by

$$h(x) = \frac{F(x) - F(a)}{G(x) - G(a)}$$

with

$$F(x) = f(x) - f(a) - (x - a) f'(a) \quad \text{and} \quad G(x) = (x - a)^2.$$

Note that $F(a) = G(a) = 0$. Since $G'(x) = 2(x - a) \neq 0$ for $x \neq a$, there must be a point ξ between a and x such that

$$h(x) = \frac{F'(\xi)}{G'(\xi)} = \frac{f'(\xi) - f'(a)}{2(\xi - a)}. \tag{4.8}$$

It doesn't really matter whether $x > a$ or $x < a$. In the first case we use the interval $[a, x]$ and in the second we use $[x, a]$. We cover both possibilities by simply saying that ξ is between a and x.

When f' is differentiable at the point a, we can remove the discontinuity in h at a by defining

$$h(a) = \lim_{x \to a} h(x) = \lim_{\xi \to a} \frac{f'(\xi) - f'(a)}{2(\xi - a)} = \frac{1}{2} f''(a).$$

The quantity $f''(a)$ is called the **second derivative** of f at a, and it represents the value at a of the derivative of f'. Returning to (4.8), when f' is differentiable everywhere between a and x, we can apply the mean value theorem to obtain

$$h(x) = \frac{1}{2} f''(\xi^*)$$

for some point ξ^* between a and ξ and therefore between a and x.

Now let's assemble all these ideas in a formal theorem, just to make sure we include all the needed hypotheses. We'll use it in Chapter 6 to analyze a limit that is sometimes used to define the natural exponential function, and we'll use it again in Chapter 9 to analyze Newton's method for solving nonlinear equations.

THEOREM 4.1: *Suppose that the numerical function f is continuous on $[\alpha, \beta]$ and differentiable in (α, β) and that f' is differentiable at the point $a \in (\alpha, \beta)$. Then there is a continuous function $h : [\alpha, \beta] \to \mathbf{R}$ with $h(a) = \frac{1}{2}f''(a)$ such that*

$$f(x) = f(a) + (x - a) f'(a) + (x - a)^2 h(x) \quad \text{for all } x \in [\alpha, \beta].$$

With the additional assumption that f' is differentiable everywhere in (α, β), for each $x \neq a$ in $[\alpha, \beta]$ there is a point ξ between a and x satisfying

$$f(x) = f(a) + (x - a) f'(a) + \frac{1}{2}(x - a)^2 f''(\xi).$$

Assuming only the existence of $f''(a)$ gives us a qualitative appraisal of the quadratic approximation

$$f(a) + (x - a) f'(a) + \frac{1}{2}(x - a)^2 f''(a)$$

to $f(x)$ near $x = a$. This approximation differs from $f(x)$ by an amount that becomes arbitrarily small in comparison to $(x - a)^2$ as x approaches a. When $f''(x)$ exists throughout (α, β), we have a convenient quantitative expression for this difference in the form

$$\frac{1}{2}(x - a)^2 \left[f''(a) - f''(\xi) \right]$$

for some ξ between a and x.

While quadratic approximations are often used in quantitative work, the theorem also has important qualitative implications. The continuity of h at a shows that $h(x)$ and $f''(a)$ must have the same sign at points near a, so the graph of f is above the tangent line where $f'' > 0$ and below the tangent line where $f'' < 0$. In particular, where the tangent line is horizontal the sign of the second derivative determines whether the point of tangency represents a relative maximum or relative minimum; this is the familiar **second derivative test** for relative extrema.

There's a more general version of Theorem 4.1 that deals with approximations using polynomials of higher degree with hypotheses involving derivatives of higher order. It's known as **Taylor's theorem**, and we discuss it in Chapter 8. Such higher-order approximations are potentially more accurate but somewhat harder to work with.

Figure 4.1 is a graph of $f(x) = 1/(x^2 + 1)$ along with its linearization and quadratic approximation around $x = 1$, with the graph of f shown as a heavy curve. Near $x = 1$ the quadratic approximation produces a curve that almost exactly matches the original graph, making it a noticeably better approximation than the linearization; however, at more distant points the quadratic approximation may be less accurate.

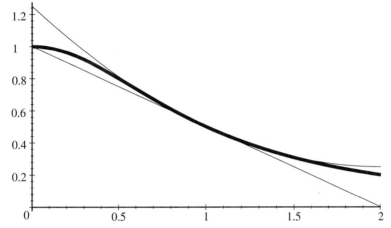

Figure 4.1   Linearization and quadratic approximation for $1/(x^2 + 1)$.

EXERCISES

14. Prove that if $f''(a)$ exists and is not zero, then there is an open interval I about a where the graph of f falls inside the graph of the parabola

$$f(a) + (x - a)f'(a) + \frac{1}{4}(x - a)^2 f''(a).$$

15. The quadratic approximation to $\sqrt{x}$ near $x = 4$ is

$$Q(x) = 2 + \frac{1}{4}(x - 4) - \frac{1}{64}(x - 4)^2.$$

Find a representation for $\sqrt{x} - Q(x)$, and use it to bound the error in the quadratic approximation by a small constant multiple of $(x - 4)^2$ for $x \in [3, 5]$.

5   HIGHER DERIVATIVES

Since the derivative of a numerical function is another numerical function, as long as we continue to produce differentiable functions we can repeat the process of differentiation indefinitely. Instead of using additional primes to indicate higher derivatives, we write $f^{(n)}$ for the function obtained from f by differentiating it n times. To unify our notation, we agree that $f^{(0)}$ is simply f. Leibniz's notation for the nth derivative is $\frac{d^n y}{dx^n}$; we'll occasionally find this notation more convenient. Mathematicians call $C^n(I)$ the class of all numerical functions such that $f^{(n)}$ is continuous on I. Since every point where a function is differentiable is a point of continuity for the function, we see that the classes $C^n(I)$ decrease with n; that is, $C^{n+1}(I) \subset C^n(I)$ for all natural numbers n. Functions that are in $C^n(I)$ for every natural number n are said to be **infinitely differentiable** on I; $C^\infty(I)$ represents this class of functions. Every polynomial function is infinitely differentiable at all points, but the class $C^\infty(I)$ includes many other functions even when I is all of $\mathbf{R}$.

Higher derivatives are useful in trying to find the roots of polynomials, a problem that comes up in many contexts. We often try to find roots of a polynomial p by simply seeing whether or not $p(x) = 0$ for various values of x. Repeated roots, however, present special difficulties. Clearly, something else besides evaluating $p(x)$ must be done to identify a repeated root. Derivatives of higher order can come to the rescue, and it's easy to understand why.

When $x = a$ is a root of the polynomial $p(x)$, we call it a **k-fold root** if $k > 0$ and there is a polynomial $q(x)$ such that

$$p(x) = (x - a)^k q(x) \quad \text{with } q(a) \neq 0.$$

In this case, the product rule tells us that

$$\begin{aligned} p'(x) &= k(x - a)^{k-1} q(x) + (x - a)^k q'(x) \\ &= (x - a)^{k-1} \left[kq(x) + (x - a) q'(x) \right] \\ &= (x - a)^{k-1} q_1(x), \end{aligned}$$

where

$$q_1(x) = kq(x) + (x - a) q'(x).$$

Note that q_1 is also a polynomial, and if $k > 1$ then $p'(x)$ has a $(k - 1)$-fold root since

$$q_1(a) = kq(a) \neq 0.$$

Taking additional derivatives shows us that for $n \leq k$,

$$p^{(n)}(x) = (x - a)^{k-n} q_n(x),$$

where $q_n(x)$ is a polynomial with $q_n(a) \neq 0$.

So when p has a k-fold root at a, then $p^{(n)}(a) = 0$ for $n < k$ but not for $n = k$. That's a theorem and we've just proved it; we'll state it next.

THEOREM 5.1: *Let $p(x)$ be a polynomial and let k be a positive integer. Then $p(x)$ has a k-fold root at $x = a$ if and only if*

$$p(a) = p'(a) = p''(a) = \cdots = p^{(k-1)}(a) = 0 \quad \text{with } p^{(k)}(a) \neq 0.$$

As we use more sophisticated applications for derivatives, we encounter situations where we need to take an nth derivative of the product of two functions. Fortunately, there's a fairly simple formula that covers such cases, known as **Leibniz's rule**. Here's a statement of it.

THEOREM 5.2: *Suppose that x is a point at which both f and g can be differentiated n times. Then their product h can also be differentiated n times at x, and*

$$h^{(n)}(x) = \sum_{k=0}^{n} \frac{n!}{k!\,(n-k)!} f^{(n-k)}(x)\, g^{(k)}(x).$$

□ *Proof:* We prove the theorem using mathematical induction. The case $n = 1$ is simple; it says that when f and g are differentiable at x their product h is also differentiable, and

$$h^{(1)}(x) = \frac{1!}{0!1!} f^{(1)}(x)\, g^{(0)}(x) + \frac{1!}{1!0!} f^{(0)}(x)\, g^{(1)}(x).$$

That's just a messy statement of the familiar product rule, so we know that Leibniz's rule is true when $n = 1$.

Now we assume that N is a positive integer such that Leibniz's rule is true for $n = N$, and prove that it must also be true for $n = N + 1$. The hypotheses for this case are that f and g may each be differentiated $N + 1$ times at x. Then they're also differentiable N times at x, and since we've assumed that Leibniz's rule is true for $n = N$,

$$h^{(N)}(x) = \sum_{k=0}^{N} \frac{N!}{k!\,(N-k)!} f^{(N-k)}(x)\, g^{(k)}(x).$$

When we examine the orders of the derivatives of f and g appearing in this sum, we note that both k and $N - k$ are less than $N + 1$. So by the product rule, each term $f^{(N-k)}(x)\, g^{(k)}(x)$ is also differentiable, and hence $h^{(N+1)}(x)$ is given by

$$\sum_{k=0}^{N} \frac{N!}{k!\,(N-k)!} \left[f^{(N+1-k)}(x)\, g^{(k)}(x) + f^{(N-k)}(x)\, g^{(k+1)}(x) \right].$$

To complete the proof, we need to show that this sum can be regrouped as

$$\sum_{k=0}^{N+1} \frac{(N+1)!}{k!\,(N+1-k)!}\, f^{(N+1-k)}(x)\, g^{(k)}(x).$$

That's possible with a diligent application of algebra, but there's an easier way that's also more informative. According to the binomial theorem,

$$(A+B)^N = \sum_{k=0}^{N} \frac{N!}{k!\,(N-k)!}\, A^{N-k} B^k,$$

so

$$(A+B)^{N+1} = (A+B) \sum_{k=0}^{N} \frac{N!}{k!\,(N-k)!}\, A^{N-k} B^k$$

$$= \sum_{k=0}^{N} \frac{N!}{k!\,(N-k)!} \left(A^{N+1-k} B^k + A^{N-k} B^{k+1} \right).$$

However, we know that can be regrouped to give

$$(A+B)^{N+1} = \sum_{k=0}^{N+1} \frac{(N+1)!}{k!\,(N+1-k)!}\, A^{N+1-k} B^k.$$

The algebra involved must be exactly the same for both cases, so no further proof is necessary.   ■

As we use Leibniz's rule, it is often helpful to remember that the numerical coefficients are just the binomial coefficients, and they can be calculated easily from Pascal's triangle. Indeed, Pascal's triangle is nothing more than an efficient bookkeeping device for the cumulative effect of the algebraic regrouping operation involved.

EXERCISES

16. Show that if $g \in C^n(I)$ and $h(x) = x g(x)$, then $h \in C^n(I)$ and

$$h^{(n)}(x) = x g^{(n)}(x) + n g^{(n-1)}(x).$$

(While this is a straightforward application of Leibniz's rule, it is also easy to establish by induction.)

17. Show that if $g \in C^n(I)$ and $h(x) = x^2 g(x)$, then $h \in C^n(I)$ and

$$h^{(n)}(x) = x^2 g^{(n)}(x) + 2nx g^{(n-1)}(x) + n(n-1) g^{(n-2)}(x),$$

with the last term understood to be 0 when $n = 1$. Either Leibniz's rule or Exercise 16 can be used.

6 INVERSE FUNCTIONS

We often deal with pairs of functions having the property that one undoes the other. For example, if f and ϕ are numerical functions defined by the formulas

$$f(x) = 2x + 1 \quad \text{and} \quad \phi(x) = \frac{1}{2}(x - 1),$$

then it's easy to calculate that

$$f(\phi(x)) = 2\left[\frac{1}{2}(x - 1)\right] + 1 = x$$

as well as

$$\phi(f(x)) = \frac{1}{2}([2x + 1] - 1) = x.$$

We describe this situation by saying that f and ϕ are **inverse functions**; we may either identify f as the inverse of ϕ or ϕ as the inverse of f.

In the example above, the formula for ϕ was calculated from the formula for f. The procedure was to solve the equation

$$x = f(y)$$

for y in terms of x. The formula obtained for y was then used as the formula for $\phi(x)$. This example was especially simple because we could solve explicitly for y using basic algebra. For each x we found exactly one y, and since the domain of f was all of $\mathbf{R}$ we knew that the y we found was acceptable. In other cases we may not be so lucky; we may not know how to perform the operations involved, and we may not be able to determine the domains of the functions so easily. If we're going to reason with inverse functions correctly, we'll need a more precise definition to deal with the general case.

DEFINITION 6.1: Given two functions $f : A \to B$ and $\phi : B \to A$, we say that f and ϕ are inverse functions if

$$\phi(f(x)) = x \text{ for all } x \in A \quad \text{and} \quad f(\phi(y)) = y \text{ for all } y \in B.$$

This is one context where we need to pay careful attention to all the formalism in the language of functions. The question of whether f has an inverse depends on both the sets A and B as well as the formula for f. For example, if we define

$$f : [0, \infty) \to [0, \infty) \text{ by } f(x) = x^2$$

and

$$\phi : [0, \infty) \to [0, \infty) \text{ by } \phi(x) = \sqrt{x},$$

then f and ϕ are inverse functions. But it we enlarge either the domain or the range of f to include some negative numbers as well, then f no longer has an inverse.

It's important to note that the definition of inverse functions involves two equations; it's possible to satisfy one but not the other. For example, suppose $A = [0, 2]$, $B = [0, 1]$, and we define $f : A \to B$ by $f(x) = |x - 1|$ and $\phi : B \to A$ by $\phi(y) = 1 + y$. While

$$f(\phi(y)) = |y| = y \text{ for all } y \in B$$

we note that

$$\phi(f(x)) = 1 + |x - 1|.$$

This only agrees with x when $1 \le x \le 2$, not for all $x \in A$.

You may have noticed that something seems wrong with the example we just considered. While $\phi : B \to A$ we see that $\{\phi(x) : x \in B\}$, the set of values of ϕ, is only $[1, 2]$ instead of being all of A. That's a feature common to any example we might create. When $f : A \to B$ and $\phi : B \to A$ with $f(\phi(y)) = y$ for all $y \in B$, then $\phi(f(x)) = x$ for each $x \in A$ that is a value of ϕ. That's easy to see; when $x = \phi(y)$ our assumptions tell us that

$$\phi(f(x)) = \phi(f(\phi(y))) = \phi(y) = x.$$

Inverse functions are extremely convenient to work with when we have them. For example, for each $y \in B$ we can simply say that $\phi(y)$ is the one value of $x \in A$ such that $f(x) = y$. They're often a convenient way to identify a needed solution of an equation when we can't or don't want to find an explicit formula for it. The problem is that inverse functions are elusive. Most functions don't have inverses and algebra isn't much help in identifying many of those that do. Fortunately, the methods of calculus can provide both simple ways for determining that an inverse does exist and explicit methods for working with it, whether or not we find its formula. The theorem below shows how it's done; we'll naturally refer to it as the **inverse function theorem.**

THEOREM 6.1: *Let I be an open interval, and let f be a numerical function with domain I. If f has no critical points in I, then the image of I under f is an open interval J, and f has an inverse function $\phi : J \to I$. Moreover, ϕ is differentiable, and its derivative satisfies the equation*

$$\phi'(y) = \frac{1}{f'(x)} \quad \text{for all } x \in I, \ y \in J \text{ with } y = f(x).$$

The hypothesis that f has no critical points is a powerful one. Since a critical point is any x in the domain of the function where either $f'(x)$ is undefined or $f'(x) = 0$, we've assumed that f is differentiable at every point in I and its derivative is never 0. While exploring the mean value theorem, we saw that the set of values

$$\{ f'(x) : x \in I \}$$

is itself an interval when f is differentiable at all points of the open interval I, so to satisfy the hypothesis $f'(x)$ must either be positive at all points in I or be negative everywhere.

When we know that f and ϕ are inverse functions and that both are differentiable, the relationship between their derivatives is an easy consequence of the chain rule. Since $\phi(f(x)) = x$ for all $x \in I$, we know that

$$\frac{d}{dx}(x) = \frac{d}{dx}\{\phi(f(x))\} = \phi'(f(x))f'(x),$$

and we just substitute y for $f(x)$ and recognize that the derivative of x is 1 to complete the derivation. But the theorem says more than that; we don't need to know in advance that both functions are differentiable to use it.

□ *Proof:* Now let's prove the theorem. The first assertion is that the set of values

$$J = \{ f(x) : x \in I \}$$

defines an open interval. Since every differentiable function is continuous, the intermediate value theorem guarantees that J must be an interval. Since f has no critical points, none of its values can be extreme values. This proves that J has neither a least or a greatest value, so J is an open interval.

The next thing to prove is that f has an inverse function. There's only one way possible to define it: for $y \in J$ we must define $\phi(y)$ to be the one $x \in I$ such that $f(x) = y$. Since J was defined to be the set of values of f, we know there is at least one such x. If there were two, then Rolle's theorem would require that there be a critical point for f between them. But we've assumed that there are no critical points in I, so for each $y \in J$ there is indeed exactly one $x \in I$ with $f(x) = y$.

Instead of next proving that ϕ is differentiable, we'll show that it's continuous on J; this turns out to be a necessary step in the argument. Suppose that $b \in J$ and we're given an $\varepsilon > 0$. We need to find a $\delta > 0$ such that

$$|\phi(y) - \phi(b)| < \varepsilon \quad \text{for every } y \in (b - \delta, b + \delta) \cap J.$$

Call $\phi(b) = a \in I$. Since I is an open interval, it doesn't hurt to assume that our given ε is small enough that both $a - \varepsilon$ and $a + \varepsilon$ are in I. Then

$$\{f(x) : a - \varepsilon < x < a + \varepsilon\}$$

is an open interval for exactly the same reasons that J is, and we know that b is a point in it. We simply pick our δ small enough that

$$(b - \delta, b + \delta) \subset \{f(x) : a - \varepsilon < x < a + \varepsilon\},$$

and then every $y \in (b - \delta, b + \delta)$ will satisfy

$$\phi(y) \in (a - \varepsilon, a + \varepsilon).$$

Now we're ready to show that ϕ is differentiable. In the process, we'll verify the formula for its derivative. We prove differentiability one point at a time by analyzing $\phi(y)$ near each fixed $b \in J$; the continuity of ϕ plays an important role. Once again we'll call

$$a = \phi(b) \quad \text{and} \quad x = \phi(y).$$

Since f is assumed to be differentiable at a, there is a numerical function g defined on I that is continuous at a and satisfies

$$f(x) = f(a) + (x - a) g(x). \tag{4.9}$$

We can rewrite equation (4.9) as

$$y = b + [\phi(y) - \phi(b)] g(\phi(y)),$$

and so as long as $g(\phi(y)) \neq 0$ we have

$$\phi(y) = \phi(b) + \frac{y - b}{g(\phi(y))}.$$

In fact, the function g defined by equation (4.9) is never 0 on I, because all its values are values of f' on I. That's true by definition when $x = a$ and it's a consequence of the mean value theorem for all other $x \in I$. Hence the last equation is true for every $y \in J$, and the formula

$$G(y) = \frac{1}{g(\phi(y))}$$

defines a numerical function G on I. Since ϕ is continuous at b and g is continuous at a, we see that G is continuous at b. Hence ϕ is differentiable at b, with

$$\phi'(b) = G(b) = \frac{1}{g(\phi(b))} = \frac{1}{g(a)}.$$

That completes our proof of the inverse function theorem.   ■

EXERCISES

18. Prove that if $f : A \to B$ and $\phi : B \to A$ with $f(\phi(y)) = y$ for all $y \in B$, then f and ϕ are inverse functions unless there are $x_1, x_2 \in A$ with $x_1 \neq x_2$ but $f(x_1) = f(x_2)$.

19. Show that $f(x) = x^3 + x$ defines a function on all of **R** that has a differentiable inverse. Find the domain of the inverse function ϕ, and calculate $\phi'(2)$.

20. Suppose $f(x) = \frac{1}{2}\left[-b + \sqrt{b^2 - 4c + 4x}\right]$, where b and c are fixed real numbers, and the domain of f is understood to be the set of all x at which the given expression defines a differentiable function. Show that f satisfies the hypotheses of Theorem 6.1, and find an explicit formula for its inverse ϕ. Calculate f' and ϕ' from their formulas, and show they satisfy the conclusion of Theorem 6.1.

21. Suppose that f satisfies all the hypotheses of Theorem 6.1, and that $b = f(a)$ is a point in the domain of the inverse function ϕ. Prove that if f' is differentiable at a, then ϕ' is differentiable at b, and find a formula for $\phi''(b)$.

22. Even though $f(x) = x^3$ has a critical point at 0, f still has an inverse with domain and range all of **R**; the inverse function is $x^{1/3}$. How can Theorem 6.1 be used to show that $x^{1/3}$ is differentiable at all points except the origin?

7   IMPLICIT FUNCTIONS AND IMPLICIT DIFFERENTIATION

Sometimes we need to work with the graph of an equation

$$F(x, y) = C$$

that we can't solve either for y as a function of x or for x as a function of y. Plotting such graphs is difficult even for computers. Instead of finding points by substituting values for one coordinate and calculating the corresponding value of the other, it may be necessary to substitute possible values for both coordinates and see whether the equation is satisfied. To aid in the study of such graphs, most introductory calculus texts present a technique known as **implicit differentiation**. To get started, we need the coordinates of a point (a, b) on the graph. Briefly, the method is to assume that near the known point, the equation defines y as a differentiable function $\phi(x)$ on an open interval I about a, in such a way that

$$F(x, \phi(x)) = C \quad \text{for all } x \in I.$$

Then $\phi'(x)$ can be computed by differentiating this equation with respect to x and solving for $\phi'(x)$ in terms of x and $\phi(x)$. The equation to be solved should be linear in the unknown $\phi'(x)$, and the procedure is supposed to work as long as the coefficient of $\phi'(x)$ is nonzero. Setting $x = a$ and $\phi(x) = b$ in the formula for $\phi'(x)$ then gives the slope of the tangent line at (a, b).

Often the introduction of this method brings a great deal of confusion, partly because the underlying concepts haven't even been given names, let alone clear descriptions. Lucid explanations of the method require some understanding of functions of several variables and of the basic methods for dealing with them, and that understanding is generally missing in basic calculus courses. The study of functions of more than one variable gets much more complicated than what we've been doing up to now. However, the part of the theory needed to understand implicit differentiation is not especially complicated; we'll develop it below.

Typically $F(x, y)$ is a simple algebraic expression involving two unknowns x and y. But in principle it can be anything that assigns a numerical value to each point (x, y) in some region of the coordinate plane, as long as the value assigned depends only on the location of the point. For (a, b) in the region of the plane where F is defined, we say that F is continuous at (a, b) if for every given $\varepsilon > 0$, there is a number $\delta > 0$ such that every point (x, y) in that region with both $|x - a| < \delta$ and $|y - b| < \delta$ also satisfies $|F(x, y) - F(a, b)| < \varepsilon$. This is exactly the same sort of condition we developed for doing arithmetic with numbers known only approximately, and we can use it in exactly the same way. It closely parallels the concept of continuity we developed for functions defined on subsets of the real line; instead of requiring that a condition hold for all x in the interval of length 2δ centered at a, we require it to be true for all (x, y) inside the square of side 2δ centered at (a, b).

Calculus teaches us to study functions by looking carefully at the way their values change, and to analyze the changes in a function of two variables it helps to think of the variables as changing one at a time. That is, to see how $F(x, y)$ differs from $F(a, b)$, we first examine the difference between $F(x, b)$ and $F(a, b)$ and then look at the difference between $F(x, y)$ and $F(x, b)$. By temporarily regarding one of the variables as frozen, we analyze the change in one function of two variables in terms of the changes in two functions of one variable. In particular, the **partial derivative** of $F(x, y)$ with respect to x is found by treating y as a constant and finding the derivative of the corresponding function of x. Similarly, the partial derivative with respect to y is found by treating x as a constant and finding the derivative of the corresponding function of y. That's the general idea, but there are enough technical details involved that we need to give more precise statements of what we mean.

DEFINITION 7.1: Given a function F of two variables, suppose that (a, b) is a point such that $F(x, b)$ is defined for all x in an open interval I_1 containing a, and also that $F(a, y)$ is defined for all y in an open interval I_2 containing b. We define

$$F_1(a, b) = \lim_{x \to a} \frac{F(x, b) - F(a, b)}{x - a}$$

and
$$F_2(a, b) = \lim_{y \to b} \frac{F(a, y) - F(a, b)}{x - a}$$

provided these limits exist. The quantities $F_1(a, b)$ and $F_2(a, b)$ are called partial derivatives of F; we think of them as functions of the point (a, b).

Just as there are several different notations commonly used for derivatives, there are other ways to indicate partial derivatives; the notation we've chosen seems best for our purposes. We'll often substitute new letters in place of variables, so we've chosen to keep track of the arguments of F as the first or second variable, rather than by the letter used. That is, we calculate F_1 by differentiating $F(x, y)$ with respect to x, the first variable, and we calculate F_2 by differentiating with respect to the second. Then x and y can be replaced by numerical expressions as needed.

The condition that allows us to approximate $F(x, y)$ near (a, b) by a linear function of two variables is again called **differentiability**. We say that F is differentiable at (a, b) if its domain includes all points inside some square

$$S_\delta = \{(x, y) : |x - a| < \delta \text{ and } |y - b| < \delta\}$$

centered at (a, b) and if there are functions $G_1, G_2 : S_\delta \to \mathbf{R}$ that are continuous at (a, b) and satisfy

$$F(x, y) = F(a, b) + (x - a) G_1(x, y) + (y - a) G_2(x, y) \qquad (4.10)$$

at all points of S_δ. (Here we are not assuming that G_1 and G_2 are partial derivatives of a function G; the subscripts are used to link them to F_1 and F_2.) Since (4.10) leads to

$$F(x, b) = F(a, b) + (x - a) G_1(x, b)$$
and $\qquad F(a, y) = F(a, b) + (y - b) G_1(a, y),$

differentiability at (a, b) implies the existence of both partial derivatives at that point, with $F_1(a, b) = G_1(a, b)$ and $F_2(a, b) = G_2(a, b)$. However, the existence of $F_1(a, b)$ and $F_2(a, b)$ depends only on the values of F along horizontal and vertical lines through (a, b), while differentiability at (a, b) involves the values of $F(x, y)$ at all points in S_δ. But even though it is possible for both partial derivatives to exist at a point where F is not differentiable, there is still a way to use partial derivatives to check for differentiability. The theorem below tells how to do it.

THEOREM 7.1: *Suppose that the function F has partial derivatives F_1 and F_2 defined at every point in a square S_δ centered at (a, b). If F_1 and F_2 are continuous at (a, b), then F is differentiable at (a, b).*

□ *Proof:* While equation (4.10) doesn't determine the functions G_1 and G_2, there's a simple way to define them to make it be satisfied. Of course, we want $G_1(a, b) = F_1(a, b)$ and $G_2(a, b) = F_2(a, b)$, so for

$(x, y) \neq (a, b)$ we define

$$G_1 (x, y) = F_1 (a, b) + \frac{(x - a) E (x, y)}{(x - a)^2 + (y - b)^2} \tag{4.11}$$

and

$$G_2 (x, y) = F_2 (a, b) + \frac{(y - b) E (x, y)}{(x - a)^2 + (y - b)^2}, \tag{4.12}$$

where

$$E (x, y) = F (x, y) - F (a, b) - (x - a) F_1 (a, b) - (y - b) F_2 (a, b).$$

Direct substitution shows that (4.10) is satisfied, and we may note that this definition of G_1 and G_2 makes them continuous at all points other than (a, b) where F is continuous. To prove continuity at (a, b), we need to analyze E; that's where the assumptions about F_1 and F_2 are needed.

For each fixed $y \in (b - \delta, b + \delta)$, we can regard $F (x, y)$ as a differentiable function of x on $(a - \delta, a + \delta)$. So for each x in this interval the mean value theorem guarantees that there is a number ξ between a and x with

$$F (x, y) = F (a, y) + (x - a) F_1 (\xi, y).$$

Also, we can regard $F (a, y)$ as a differentiable function of y on $(b - \delta, b + \delta)$. So for each y in that interval there is a number η between b and y such that

$$F (a, y) = F (a, b) + (y - b) F_2 (a, \eta).$$

Combining these equations shows

$$F (x, y) = F (a, b) + (x - a) F_1 (\xi, y) + (y - b) F_2 (a, \eta),$$

and therefore $E (x, y)$ is given by

$$(x - a) [F_1 (\xi, y) - F_1 (a, b)] + (y - b) [F_2 (a, \eta) - F_2 (a, b)].$$

Note that if we restrict (x, y) to a smaller square centered at (a, b), then both (ξ, y) and (a, η) will be in this same square. Hence the continuity of F_1 and F_2 at (a, b) shows that for any given $\varepsilon > 0$, there is a $\delta' > 0$ such that the coefficients of $(x - a)$ and $(y - b)$ in the formula for E will both be in $(-\varepsilon/2, \varepsilon/2)$ whenever both $|x - a| < \delta'$ and $|y - b| < \delta'$. Hence for all such (x, y) we have

$$|E (x, y)| < \frac{\varepsilon}{2} |x - a| + \frac{\varepsilon}{2} |y - b|$$

$$< \varepsilon \sqrt{(x - a)^2 + (y - b)^2}. \tag{4.13}$$

Since both $|x - a|$ and $|y - b|$ are bounded by $\sqrt{(x - a)^2 + (y - b)^2}$, using (4.13) in (4.11) and (4.12) shows both

$$|G_1(x, y) - F_1(a, b)| < \varepsilon \quad \text{and} \quad |G_2(x, y) - F_2(a, b)| < \varepsilon$$

for all (x, y) with both $|x - a| < \delta'$ and $|y - b| < \delta'$, proving that G_1 and G_2 are continuous at (a, b). That completes the proof of the theorem. ∎

Now let's take another look at the method of implicit differentiation. We begin by assuming that F is differentiable at (a, b) and that $\phi(a) = b$ with ϕ differentiable at a. Define

$$h(x) = F(x, \phi(x)).$$

We'll show that h is differentiable at a and calculate $h'(a)$. Since ϕ continuous at a, there must be an open interval $I = (a - \delta, a + \delta)$ on which all the values of $\phi(x)$ are close enough to b to make

$$h(x) = F(a, b) + (x - a)G_1(x, \phi(x)) + (\phi(x) - b)G_2(x, \phi(x)),$$

with G_1 and G_2 both continuous at (a, b). The continuity of ϕ at a guarantees that both $G_1(x, \phi(x))$ and $G_2(x, \phi(x))$ define functions of x that are continuous at a, and the differentiability of ϕ at a lets us write

$$\phi(x) - b = (x - a)g(x) \quad \text{for all } x \in I,$$

with g continuous at a. So we find

$$h(x) = h(a) + (x - a)G_1(x, \phi(x)) + (x - a)g(x)G_2(x, \phi(x)).$$

We've proved that h is differentiable at a with

$$\begin{aligned} h'(a) &= G_1(a, \phi(a)) + g(a)G_2(a, \phi(a)) \\ &= F_1(a, b) + \phi'(a)F_2(a, b). \end{aligned}$$

Of course, we can also find $h'(a)$ by substituting $y = \phi(x)$ in the formula for F and just using the chain rule along with whatever other differentiation formulas are appropriate, but it must also be possible to group the results in the form above.

For implicit differentiation, we also assume that the graph of ϕ over I coincides with the graph of the equation $F(x, y) = C$, so of course $h(x) = C$ for all $x \in I$ and $h'(a) = 0$. As long as $F_2(a, b) \neq 0$, we can calculate $\phi'(a)$ by simply solving

$$F_1(a, b) + \phi'(a)F_2(a, b) = 0.$$

That's how implicit differentiation is supposed to work.

Note that we made several significant assumptions along the way. While the differentiability of F at the point (a, b) on the graph of $F(x, y) = C$ can be verified by analyzing the function F used to form the equation, the assumptions about ϕ are another matter entirely. Of course, when the graph of $F(x, y) = C$ is a familiar object and it's known to be a smooth curve near (a, b), those assumptions ought to be satisfied. But we'd especially like to use the method when we don't know much about the graph. That's where the next result, commonly known as the **implicit function theorem**, comes in. The hypotheses we need include the hypotheses of Theorem 7.1, not just its conclusion.

THEOREM 7.2: *Let (a, b) be a point on the graph of $F(x, y) = C$, where C is a given constant, and suppose that both partial derivatives F_1 and F_2 of F are defined at all points in a square S centered at (a, b). If F_1 and F_2 are continuous at (a, b) with $F_2(a, b) \neq 0$, then there is a numerical function ϕ that is differentiable at a and satisfies $F(x, \phi(x)) = C$ for all x in its domain. Also, there is a square S_δ centered at (a, b) in which the only points (x, y) on the graph of $F(x, y) = C$ are those on the graph of ϕ. The derivative of ϕ at a is given by*

$$\phi'(a) = -F_1(a, b) / F_2(a, b).$$

☐ *Proof:* Since F satisfies the hypotheses of Theorem 7.1, for $(x, y) \in S$ we can write

$$F(x, y) = C + (x - a) G_1(x, y) + (y - b) G_2(x, y),$$

with G_1 and G_2 continuous at (a, b) and $G_2(a, b) = F_2(a, b) \neq 0$. By continuity, we can find a smaller square S_δ centered at (a, b) on which F_2 and G_2 keep the same sign, and the ratio G_1 / G_2 satisfies the bound

$$\left| \frac{G_1(x, y)}{G_2(x, y)} \right| \leq 1 + \left| \frac{F_1(a, b)}{F_2(a, b)} \right| = M.$$

Let's call I the interval $(a - \delta, a + \delta)$ and J the interval $(b - \delta, b + \delta)$.

For any $c, d \in J$ with $c < d$ and for any fixed $x \in I$, the function f defined by $f(y) = F(x, y)$ is continuous on $[c, d]$ and has no critical points in that interval, since $f'(y) = F_2(x, y) \neq 0$. Thus Rolle's theorem shows that $F(x, c) \neq F(x, d)$. So for any $x \in I$, there can never be more than one y with $(x, y) \in S_\delta$ and $F(x, y) = C$. If we define A to be the set of $x \in I$ for which there is such a y and call $\phi(x)$ that unique value of y, then ϕ is a numerical function with domain A, $\phi(a) = b$, and the points

(x, y) in S_δ on the graph of $F(x, y) = C$ are precisely those on the graph of ϕ.

The next step is to prove that A includes an open interval containing a; that's an important part of proving that ϕ is differentiable at a. We pick $r > 0$ with $b - r$ and $b + r$ in J, and then note that C must be strictly between $F(a, b - r)$ and $F(a, b + r)$. Indeed, we have

$$[F(a, b - r) - C][F(a, b + r) - C] = -r^2 G_2(a, b - r) G_2(a, b + r),$$

which is negative since G_2 keeps a fixed sign in S_δ. Since $F(x, b - r)$ and $F(x, b + r)$ are functions of x that are continuous at a, there must be an interval $(a - \varepsilon, a + \varepsilon)$ on which C is always between $F(x, b - r)$ and $F(x, b + r)$. At each such x, the function f defined by

$$f(y) = F(x, y)$$

is continuous on the interval $[b - r, b + r]$ and C is between the values of f at the endpoints. So the intermediate value theorem guarantees there is a y in the interval with $f(y) = C$. But then $(x, y) \in S_\delta$ with $F(x, y) = C$, and therefore x is in A, the domain of ϕ.

We complete the proof by showing that ϕ is differentiable at a. For each $x \in A$ we have

$$\begin{aligned}
0 &= F(x, \phi(x)) - F(a, b) \\
&= (x - a) G_1(x, \phi(x)) + (\phi(x) - \phi(a)) G_2(x, \phi(x)),
\end{aligned}$$

and since $G_2(x, \phi(x)) \neq 0$ for $x \in A$, we can rewrite this as

$$\phi(x) - \phi(a) = -(x - a) \frac{G_1(x, \phi(x))}{G_2(x, \phi(x))}.$$

Our choice of S_δ guarantees that

$$|\phi(x) - \phi(a)| = |x - a| \left| \frac{G_1(x, \phi(x))}{G_2(x, \phi(x))} \right| \leq M |x - a|,$$

so obviously ϕ is continuous at a. Consequently,

$$g(x) = -\frac{G_1(x, \phi(x))}{G_2(x, \phi(x))}$$

defines a function on $(a - \varepsilon, a + \varepsilon)$ that is continuous at a, and for all x in that interval we have

$$\phi(x) = \phi(a) + (x - a) g(x).$$

Hence ϕ is differentiable at a, with

$$\phi'(a) = g(a) = -\frac{G_1(a, \phi(a))}{G_2(a, \phi(a))} = -\frac{F_1(a, b)}{F_2(a, b)}.$$

That completes our proof of the implicit function theorem. ■

Typically, the hypotheses about F in the implicit function theorem are satisfied at all points in a square centered at (a, b), not just at the single point, and then we have

$$\phi'(x) = -\frac{F_1(x, \phi(x))}{F_2(x, \phi(x))}$$

for all x in an open interval about a. With suitable F_1 and F_2, we can then calculate ϕ'' and successively higher derivatives of ϕ.

Generally we won't be able to find an algebraic formula for the function $\phi(x)$. Nevertheless, we can still form the linearization

$$\phi(a) + (x - a)\phi'(a) = b - (x - a)\frac{F_1(a, b)}{F_2(a, b)}$$

or perhaps form the quadratic approximation

$$\phi(a) + (x - a)\phi'(a) + \frac{1}{2}(x - a)^2 \phi''(a)$$

to approximate $\phi(x)$ near $x = a$. Such approximations are adequate for many purposes. Indeed, one of the reasons that calculus is so useful is that it gives us practical methods for working with unknown functions.

EXERCISES

23. The numerical function F defined by

$$F(x, y) = \begin{cases} \dfrac{xy}{x^2 + y^2}, & (x, y) \neq (0, 0) \\[2mm] 0, & (x, y) = (0, 0) \end{cases}$$

satisfies $F_1(0, 0) = F_2(0, 0) = 0$ since F vanishes along both coordinate axes. Show that F is neither continuous nor differentiable at $(0, 0)$.

24. For (a, b) any point except the origin, prove that the equation

$$x^2 y + y^3 = b(a^2 + b^2)$$

defines y as a differentiable function $\phi(x)$ on some open interval about a. Find the linearization of $\phi(x)$ near $x = a$.

V

THE RIEMANN INTEGRAL

O ne of the early triumphs of calculus was a general method for finding areas of plane regions bounded by curves. When f is a positive, continuous function on $[a, b]$, the area of the region bounded by the graph of f, the x-axis, and the vertical lines $x = a$ and $x = b$ is usually called the area below the graph of f. This area can be approximated by slicing the region into strips with vertical lines, estimating the areas of the strips by areas of rectangles, and then adding the areas of the strips. By reducing the width of the strips, the error in such an approximation can be made arbitrarily small, and the exact area is the limiting value of such sums as the widths of all the strips involved approach zero. The methods of calculus provide systematic ways to work with such limits and often let us determine their exact values.

This approach to the problem of finding areas is an example of **integration**, a powerful technique used in all sorts of measurements in science and mathematics. It's applicable to just about every measurement that theoretically can be performed by decomposing something into parts, measuring the parts, and adding the results. For example, volume, mass, and energy are such measurements, but pressure and temperature are not. As we look for ways to compute such measurements, the way we find areas serves as a useful model, and helps us understand the properties of the operations involved.

I AREAS AND RIEMANN SUMS

We'll study the theory of integration in the form developed by the German mathematician Bernhard Riemann in the nineteenth century. While the basic ideas had been around for some time, he is generally credited with being the first to formulate them in a way that could be applied to functions with discontinuities and still be rigorously correct. Later mathematicians found a need to extend the notion of integration to situations falling outside Riemann's theory, and several newer theories of integration are commonly encountered in higher mathematics; some of these theories are still under development.

Our first task is to define the terms we'll work with in this chapter. We'll assume that $a < b$, and that we have a numerical function f whose domain includes $[a, b]$. A **partition** of $[a, b]$ is a splitting of it into a finite number of nonoverlapping closed subintervals. The word *nonoverlapping* is carefully chosen; intervals that share interior points overlap, but closed intervals that share only an endpoint do not. Partitions of $[a, b]$ are conveniently specified in terms of a finite sequence $\{x_k\}_{k=0}^{n}$ with

$$a = x_0 < x_1 < x_2 < \cdots < x_n = b.$$

Then the partition consists of n subintervals, conveniently numbered left to right as

$$I_1 = [x_0, x_1], I_2 = [x_1, x_2], \ldots, I_n = [x_{n-1}, x_n].$$

We use the symbol $\Delta x_k = x_k - x_{k-1}$ for the length of the kth subinterval I_k. The **mesh** of the partition is our name for the largest of the lengths

$$\{\Delta x_1, \Delta x_2, ..., \Delta x_n\}.$$

We call points $\{x_1^*, x_2^*, \ldots, x_n^*\}$ **sampling points** for the partition provided that $x_k^* \in I_k$ for each $k = 1, 2, \ldots, n$. We call a sum

$$\sum_{k=1}^{n} f(x_k^*) \Delta x_k = f(x_1^*) \Delta x_1 + f(x_2^*) \Delta x_2 + \cdots + f(x_n^*) \Delta x_n$$

a **Riemann sum** for f associated with the partition $\{x_k\}_{k=0}^{n}$ as long as the points $\{x_k^*\}_{k=1}^{n}$ are sampling points for the partition. In general, the value of a Riemann sum depends on the partition and sampling points used, as well as on the function f and the interval $[a, b]$.

When f is a positive, continuous function on $[a, b]$ and we're trying to find the area below its graph, the Riemann sum formed above represents the sum of the areas of rectangles $R_1, R_2, \ldots, R_n$, with R_k having I_k for its base and $(x_k^*, f(x_k^*))$ a point in its top. Sketches suggest that the Riemann sums approach the area below the graph as we consider partitions with smaller meshes. But when f is neither positive nor continuous, the picture is much more complicated, and the Riemann sums for f need not approach a single number. To deal with this difficulty, we adopt a common strategy in mathematics—we use the operation we hope to perform to define a class of functions and then look for ways to improve our understanding of this class.

DEFINITION 1.1: For f a numerical function whose domain includes the nondegenerate closed interval $[a, b]$, we say that f is **Riemann integrable** over $[a, b]$ if there is a number, called the **integral** of f over $[a, b]$ and denoted by $\int_a^b f(x)\, dx$, such that for each given $\varepsilon > 0$, there is a $\delta > 0$ with the property that whenever $\{x_k\}_{k=0}^{n}$ is a partition of $[a, b]$ having mesh less than δ, every Riemann sum $\sum_{k=1}^{n} f(x_k^*)\, \Delta x_k$ associated with it satisfies

$$\left| \sum_{k=1}^{n} f(x_k^*)\, \Delta x_k - \int_a^b f(x)\, dx \right| < \varepsilon.$$

The process that determines the value of $\int_a^b f(x)\, dx$ is called **integration**.

That's the traditional definition of the integral. Not long ago it was given in roughly that form in just about every calculus book. It's an incredibly complicated definition, and many newer books are attempting to replace it by something simpler. Its virtue is that when we know a function to be Riemann integrable, we have ready-made schemes for approximating its integral and a powerful tool for proving things about it. But it can be extremely difficult to show that a given function satisfies the definition.

However, it's easy to verify directly that constant functions are integrable over every closed interval, with

$$\int_a^b C\, dx = C(b - a) \quad \text{for any constant } C.$$

Indeed, when $f(x) = C$ we note that $C(b-a)$ is the value of every Riemann sum associated with every partition of $[a, b]$ because

$$\sum_{k=1}^{n} f(x_k^*) \Delta x_k = \sum_{k=1}^{n} C \Delta x_k = C \left(\sum_{k=1}^{n} \Delta x_k \right) = C(b-a).$$

Usually we're not so lucky, and the values of the Riemann sums are much harder to find.

The definition is better suited to proving things about integrable functions. For example, it's easy to prove that every linear combination of integrable functions is integrable. More precisely, if f and g are Riemann integrable over $[a, b]$ and $h = \alpha f + \beta g$ with α and β constants, then h is Riemann integrable over $[a, b]$ and

$$\int_a^b [\alpha f(x) + \beta g(x)]\, dx = \alpha \int_a^b f(x)\, dx + \beta \int_a^b g(x)\, dx.$$

We prove it by showing that $\alpha \int_a^b f(x)\, dx + \beta \int_a^b g(x)\, dx$ satisfies the definition of the Riemann integral of h over $[a, b]$. We first note that for an arbitrary partition of $[a, b]$ and any choice of the sampling points, we can rewrite the Riemann sum for h in terms of the Riemann sums for f and g:

$$\sum_{k=1}^{n} h(x_k^*) \Delta x_k = \alpha \left[\sum_{k=1}^{n} f(x_k^*) \Delta x_k \right] + \beta \left[\sum_{k=1}^{n} g(x_k^*) \Delta x_k \right].$$

Given $\varepsilon > 0$, our rules for approximate arithmetic tell us that there is an $\varepsilon' > 0$ such that

$$\left| \alpha u + \beta v - \left(\alpha \int_a^b f(x)\, dx + \beta \int_a^b g(x)\, dx \right) \right| < \varepsilon$$

for all u within ε' of $\int_a^b f(x)\, dx$ and for all v within ε' of $\int_a^b g(x)\, dx$. For both f and g Riemann integrable, there must be a $\delta > 0$ such that

$$u = \sum_{k=1}^{n} f(x_k^*) \Delta x_k \quad \text{and} \quad v = \sum_{k=1}^{n} g(x_k^*) \Delta x_k$$

will satisfy these conditions whenever the mesh of $\mathcal{P}$ is smaller than δ. This will guarantee that

$$\left| \sum_{k=1}^{n} h(x_k^*) \Delta x_k - \left(\alpha \int_a^b f(x)\, dx + \beta \int_a^b g(x)\, dx \right) \right| < \varepsilon.$$

Here's another simple consequence of the definition: if f and g are Riemann integrable over $[a, b]$ with $f(x) \leq g(x)$ for all $x \in [a, b]$, then

$$\int_a^b f(x) \, dx \leq \int_a^b g(x) \, dx.$$

To prove it, we first note that if $f(x) \leq g(x)$ for all $x \in [a, b]$, then for any choice of sampling points associated with any partition of $[a, b]$ we have

$$\sum_{k=1}^n f(x_k^*) \, \Delta x_k \leq \sum_{k=1}^n g(x_k^*) \, \Delta x_k.$$

If both f and g are Riemann integrable, then for any $\varepsilon > 0$ there is a partition with mesh small enough to guarantee that the Riemann sums are within ε of the corresponding integrals. Consequently,

$$\int_a^b f(x) \, dx - \varepsilon < \int_a^b g(x) \, dx + \varepsilon$$

for every $\varepsilon > 0$, and that forces the desired inequality to be true.

The last two arguments give a hint as to why such a complicated definition developed. When we can approximate $\int_a^b f(x) \, dx$ and $\int_a^b g(x) \, dx$ separately by Riemann sums, we want to be able to approximate them simultaneously—that is, by using Riemann sums corresponding to the same partition with the same sampling points. Part of the price we pay for that convenience is that to prove integrability, we must establish a bound that is valid for every choice of sampling points with every partition of sufficiently small mesh, not just for one convenient choice. But that's more of a nuisance than an overwhelming obstacle. However, showing that the definition is satisfied requires a value for the integral, and the value can be hard to find. It wasn't at all hard in the cases we just considered; that's why they were chosen at this stage. But when rounding up the usual suspects fails to produce a likely candidate for the value of the integral, some real detective work can be required.

When f is the derivative of a function F that satisfies the hypotheses of the mean value theorem on $[a, b]$, then $F(b) - F(a)$ is the only real possibility for $\int_a^b f(x) \, dx$. With any partition, each sampling point can presumably be chosen to satisfy the conclusion of the mean value theorem on the corresponding subinterval; that is

$$F(x_k) - F(x_{k-1}) = f(x_k^*)(x_k - x_{k-1}) = f(x_k^*) \, \Delta x_k.$$

The corresponding Riemann sum is then exactly $F(b) - F(a)$. But there's much more to the story of the Riemann integral because many Riemann integrable functions are not derivatives and some derivatives are not Riemann integrable.

Many calculus texts refer to **definite integrals** and **indefinite integrals**. Our use of *integral* corresponds to the definite integral; we prefer to use **antiderivative** instead of indefinite integral. That's mainly to emphasize that the concept of integration we've defined is unrelated to the manipulations used to discover antiderivatives.

EXERCISES

1. Use the intermediate value theorem to prove that if f is positive and continuous on $[a, b]$, then there is an $x^* \in [a, b]$ such that the area below the graph of f is exactly $f(x^*)(b - a)$. Why don't we calculate this area by first finding x^*?

2. Prove that if f is positive and continuous on $[a, b]$, then for every partition $\{x_k\}_{k=0}^n$ of $[a, b]$ there are sampling points $\{x_k^*\}_{k=1}^n$ such that the corresponding Riemann sum is exactly equal to the area below the graph of f.

3. For $\{x_k\}_{k=0}^n$ an arbitrary partition of $[a, b]$ and $\bar{x}_k$ the midpoint of I_k, show that $\sum_{k=1}^n \bar{x}_k \Delta x_k = \frac{1}{2}(b^2 - a^2)$. Then use the definition to prove that x is Riemann integrable over $[a, b]$ with $\int_a^b x\, dx = \frac{1}{2}(b^2 - a^2)$. The key is to observe that

$$\sum_{k=1}^n x_k^* \Delta x_k - \frac{1}{2}(b^2 - a^2) = \sum_{k=1}^n (x_k^* - \bar{x}_k)\Delta x_k.$$

Then $|x_k^* - \bar{x}_k|$ is easy to bound in terms of the mesh of the partition.

4. Suppose that $a < b < c$, and that f is known to be Riemann integrable over $[a, b]$, $[b, c]$, and $[a, c]$. Use the definition of the integral to prove that

$$\int_a^b f(x)\, dx + \int_b^c f(x)\, dx = \int_a^c f(x)\, dx.$$

(It isn't really necessary to assume Riemann integrability over all three intervals, but without that assumption the argument is much more difficult.)

2   SIMPLIFYING THE CONDITIONS FOR INTEGRABILITY

At the end of the previous section we commented on the difficulties in showing that a given function satisfies the definition of Riemann integrability. Here we'll see that some of these difficulties can be bypassed. In particular, we'll develop a new condition that doesn't involve the value of the integral.

We begin by considering the set of all Riemann sums for f associated with a given partition of $[a, b]$. To make it easier to think of the partition as a variable, let's call

$$\mathcal{P} = \{x_k\}_{k=0}^n \quad \text{with } a = x_0 < x_1 < x_2 < \cdots < x_n = b$$

and then call $\mathcal{R}(\mathcal{P})$ the set of all Riemann sums for f associated with $\mathcal{P}$ that we can form by choosing sampling points; that is,

$$\mathcal{R}(\mathcal{P}) = \left\{ \sum_{k=1}^n f(x_k^*) \, \Delta x_k : x_1^* \in I_1, \ldots, x_n^* \in I_n \right\}.$$

Our first observation is that every function that is Riemann integrable over $[a, b]$ must be bounded on $[a, b]$. In fact, if we've got a partition $\mathcal{P}$ such that every element of $\mathcal{R}(\mathcal{P})$ is within ε of some number, that alone forces f to be bounded on $[a, b]$. Since different choices of the sampling point $x_k^* \in I_k$ can't change the resulting Riemann sum by more that 2ε, it follows that

$$\text{diam} \{f(x) : x \in I_k\} \leq 2\varepsilon/\Delta x_k \quad \text{for } k = 1, \ldots, n.$$

Thus f is bounded on each I_k and $[a, b]$ is formed from only finitely many of the subintervals I_k, so we see that f must be bounded on all of $[a, b]$; that is, we can assume that there are numbers m and M with

$$m \leq f(x) \leq M \quad \text{for all } x \in [a, b].$$

Conversely, if no such numbers exist, then we know that f cannot be Riemann integrable over $[a, b]$.

Our next observation is that we don't really need to consider arbitrary partitions with mesh less than some number; a simpler version of the definition of Riemann integrability can be given that only involves considering one partition at a time. The theorem below makes this precise.

THEOREM 2.1:   *Suppose that f is a numerical function defined on $[a, b]$, and that S is a number with the property that for each $\varepsilon > 0$, there* *is a partition $\mathcal{P}$ of $[a, b]$ with $\mathcal{R}(\mathcal{P}) \subset (S - \varepsilon, S + \varepsilon)$. Then f is Riemann integrable over $[a, b]$ and $S = \int_a^b f(x) \, dx$.*

It's obvious that the hypotheses of this theorem are satisfied whenever f satisfies our original definition of Riemann integrability, so this theorem can serve as an alternate definition once we've proved it. Proving it depends on a careful analysis of how $\mathcal{R}(\mathcal{P})$ can change as we change $\mathcal{P}$.

The simplest way to change a partition $\mathcal{P}_0$ is to reduce it to a new partition $\mathcal{P}$ by deleting one of the intermediate points x_k, so that two adjacent subintervals are collapsed to a single subinterval. In the process, we replace the two terms $f(x_k^*) \Delta x_k$ and $f(x_{k+1}^*) \Delta x_{k+1}$ by a single term corresponding to a new subinterval of length $\Delta x_k + \Delta x_{k+1}$. Since x_k^* and x_{k+1}^* are possible sampling points for the new subinterval, and the sum of the replaced terms is always between $f(x_k^*)(\Delta x_k + \Delta x_{k+1})$ and $f(x_{k+1}^*)(\Delta x_k + \Delta x_{k+1})$, we see that each Riemann sum in $\mathcal{R}(\mathcal{P}_0)$ lies between two Riemann sums in $\mathcal{R}(\mathcal{P})$. Consequently, $\mathcal{R}(\mathcal{P}_0)$ is a subset of each interval that includes $\mathcal{R}(\mathcal{P})$, and that remains true if we further change $\mathcal{P}$ by deleting additional points.

The changes to $\mathcal{P}$ that we really want to consider are not reductions; we're more interested in reversing the process by adding new intermediate points. A partition formed from $\mathcal{P}$ by inserting additional intermediate points is called a **refinement** of $\mathcal{P}$. Since we can always reduce a refinement to get back the original partition, the previous paragraph tells us how the set of Riemann sums can change when we refine the partition. In particular, whenever $\mathcal{P}'$ is a refinement of $\mathcal{P}$, we see that $\mathcal{R}(\mathcal{P}')$ must be a subset of any interval that contains $\mathcal{R}(\mathcal{P})$.

□ *Proof:* Now we prove the theorem itself. We assume that f satisfies the stated hypotheses, then prove that f satisfies the definition of Riemann integrability with $S = \int_a^b f(x) \, dx$. That is, we show that for any given $\varepsilon > 0$, there is a $\delta > 0$ such that every partition $\mathcal{P}$ with mesh less than δ must satisfy

$$\mathcal{R}(\mathcal{P}) \subset (S - \varepsilon, S + \varepsilon).$$

Our hypotheses tell us there must be a partition $\mathcal{P}_0$ with

$$\mathcal{R}(\mathcal{P}_0) \subset \left(S - \frac{1}{2}\varepsilon, S + \frac{1}{2}\varepsilon\right);$$

for the moment, consider $\mathcal{P}$ and δ as known. When we refine $\mathcal{P}$ by inserting all the points of $\mathcal{P}_0$ that aren't already there, the resulting partition $\mathcal{P}'$ is also a refinement of $\mathcal{P}_0$, so

$$\mathcal{R}\left(\mathcal{P}'\right) \subset \left(S - \frac{1}{2}\varepsilon, S + \frac{1}{2}\varepsilon\right).$$

The idea is to make sure that every Riemann sum in $\mathcal{R}\left(\mathcal{P}\right)$ is within $\frac{1}{2}\varepsilon$ of one in $\mathcal{R}\left(\mathcal{P}'\right)$. When we add a single point to $\mathcal{P}$, we need to choose only one new sampling point to produce a Riemann sum for the refined partition. So if all the values of f are in $[m, M]$, we can get a new Riemann sum that is within $(M - m)\delta$ of the old one. Extending this analysis, we see that if n is the number of points used to define $\mathcal{P}_0$, then we add no more than n new points to $\mathcal{P}$ to produce $\mathcal{P}'$. So every Riemann sum in $\mathcal{R}\left(\mathcal{P}\right)$ must be within $n\left(M - m\right)\delta$ of a Riemann sum in $\mathcal{R}\left(\mathcal{P}'\right)$. Choosing $\delta > 0$ with

$$n\left(M - m\right)\delta \leq \frac{1}{2}\varepsilon$$

completes the proof.   ∎

Our next task is to develop another characterization of the value of the Riemann integral. The example of finding the area below the graph of a positive continuous function is a useful model. Assuming only that f is a bounded function defined on $[a, b]$, for any partition $\mathcal{P}$ we may define

$$m_k = \inf\left\{f\left(x\right) : x \in I_k\right\} \quad \text{and} \quad M_k = \sup\left\{f\left(x\right) : x \in I_k\right\}$$

and use them to form

$$L\left(\mathcal{P}\right) = \sum_{k=1}^{n} m_k \Delta x_k \quad \text{and} \quad U\left(\mathcal{P}\right) = \sum_{k=1}^{n} M_k \Delta x_k,$$

commonly called the **lower sum** for f relative to $\mathcal{P}$ and the **upper sum** for f relative to $\mathcal{P}$.

In the case of a positive continuous function f, $L\left(\mathcal{P}\right)$ represents the area of an inscribed region made up of rectangles and $U\left(\mathcal{P}\right)$ represents the area of a circumscribed region, so the area under the graph must be somewhere between $L\left(\mathcal{P}\right)$ and $U\left(\mathcal{P}\right)$. In the general case, we must still have

$$\mathcal{R}\left(\mathcal{P}\right) \subset [L\left(\mathcal{P}\right), U\left(\mathcal{P}\right)]$$

because Theorem 3.4 in Chapter 1 tells us that

$$L\left(\mathcal{P}\right) = \inf \mathcal{R}\left(\mathcal{P}\right) \quad \text{and} \quad U\left(\mathcal{P}\right) = \sup \mathcal{R}\left(\mathcal{P}\right).$$

We'll see that when f is Riemann integrable, $\int_a^b f(x)\,dx$ can be defined as the one number common to all the intervals of the form $[L(\mathcal{P}), U(\mathcal{P})]$. Of course, that involves proving that there is such a number.

It's not at all obvious why there should be any number common to all the intervals in the family

$$\mathcal{I} = \{[L(\mathcal{P}), U(\mathcal{P})] : \mathcal{P} \text{ a partition of } [a, b]\},$$

but Theorem 3.2 in Chapter 1 comes to our rescue. We just need to show that every two intervals in $\mathcal{I}$ have points in common. In fact, the intersection of any two intervals in $\mathcal{I}$ must always contain a third interval in $\mathcal{I}$. Two intervals I_1 and I_2 in $\mathcal{I}$ must correspond to partitions $\mathcal{P}_1$ and $\mathcal{P}_2$ of $[a, b]$, and then for $\mathcal{P}$ any refinement of both $\mathcal{P}_1$ and $\mathcal{P}_2$ we know that $\mathcal{R}(\mathcal{P})$ is a subset of both I_1 and I_2. Consequently,

$$I = [L(\mathcal{P}), U(\mathcal{P})] \in \mathcal{I} \quad \text{with } I \subset I_1 \cap I_2.$$

When $\mathcal{I}$ includes arbitrarily short intervals, there must be exactly one point common to all the intervals in $\mathcal{I}$. That's the key to our final characterization of Riemann integrability given in the theorem below.

THEOREM 2.2: *Let f be a bounded numerical function on $[a, b]$. Then f is Riemann integrable over $[a, b]$ if for each $\varepsilon > 0$, there is a partition $\mathcal{P}$ of $[a, b]$ such that*

$$\operatorname{diam} \mathcal{R}(\mathcal{P}) = U(\mathcal{P}) - L(\mathcal{P}) < \varepsilon.$$

In that case, $\int_a^b f(x)\,dx$ is the one number common to all the intervals in the family

$$\mathcal{I} = \{[L(\mathcal{P}), U(\mathcal{P})] : \mathcal{P} \text{ a partition of } [a, b]\}.$$

□ *Proof:* For S the one number common to all the intervals in the family $\mathcal{I}$, when $U(\mathcal{P}) - L(\mathcal{P}) < \varepsilon$ we have

$$\mathcal{R}(\mathcal{P}) \subset [L(\mathcal{P}), U(\mathcal{P})] \subset (S - \varepsilon, S + \varepsilon).$$

So the conclusion follows from Theorem 2.1. It's worth noting that every Riemann integrable function must satisfy the hypotheses of Theorem 2.2. ∎

This characterization of the value of the integral makes it obvious why

$$m(b - a) \leq \int_a^b f(x)\,dx \leq M(b - a)$$

when f is Riemann integrable with $m \leq f(x) \leq M$ for all $x \in [a, b]$; just look at $L(\mathcal{P})$ and $U(\mathcal{P})$ for $\mathcal{P} = \{a, b\}$.

EXERCISES

5. For a given bounded function f on $[a, b]$ and partition $\mathcal{P}$ of $[a, b]$, show how to express the diameter of $\mathcal{R}(\mathcal{P})$ in terms of the quantities

$$D_k = \text{diam}\,\{f(x) : x \in I_k\} = \sup\{|f(x) - f(y)| : x, y \in I_k\}.$$

6. Prove that if f is Riemann integrable over $[a, b]$, then so is $|f|$. (The preceding exercise is useful.) When is

$$\int_a^b |f(x)|\, dx = \left| \int_a^b f(x)\, dx \right|?$$

7. Prove that if f is Riemann integrable over $[a, b]$, then so are f_+ and f_-, where $f = f_+ - f_-$ with

$$f_+(x) = \begin{cases} f(x), & f(x) \geq 0 \\ 0, & f(x) < 0 \end{cases}$$

and

$$f_-(x) = \begin{cases} 0, & f(x) \geq 0 \\ -f(x), & f(x) < 0. \end{cases}$$

8. Prove that if f and g are Riemann integrable over $[a, b]$, then so is fg. The simplest way is to write $f(x)g(x) - f(y)g(y)$ as

$$f(x)[g(x) - g(y)] + [f(x) - f(y)]g(y)$$

to find a useful bound for

$$\text{diam}\,\{f(x)g(x) : x \in [x_{k-1}, x_k]\}.$$

3   RECOGNIZING INTEGRABILITY

Now that we've developed Theorem 2.2, we'll use it in a way that will just about make it obsolete. We'll show that large classes of functions are Riemann integrable. Then we can establish the integrability of most of the integrable functions we encounter by inspection; we need only recognize their membership in such a class. For example, we can often recognize a function as increasing or decreasing over the entire interval of integration; such functions are called **monotonic**. We can also recognize many functions as continuous. The next two theorems state that every function in either of these classes is Riemann integrable.

THEOREM 3.1: *Every monotonic function defined on $[a, b]$ is Riemann integrable over $[a, b]$.*

☐ *Proof:* We'll use Theorem 2.2. When f is monotonic, all the values of f on $[a, b]$ will be between $f(a)$ and $f(b)$, so we know that f is bounded on $[a, b]$. If we restrict our attention to partitions involving only subintervals of equal lengths, we discover that $U(\mathcal{P}) - L(\mathcal{P})$ is quite easy to calculate.

Let's begin by assuming that f is increasing on $[a, b]$ and form a partition $\mathcal{P} = \{x_k\}_{k=0}^{n}$ by defining

$$x_k = a + \frac{k}{n}(b - a) \quad \text{for } k = 0, 1, \ldots, n.$$

Since f is increasing, we recognize that

$$m_k = \inf\{f(x) : x \in I_k\} = f(x_{k-1})$$

and

$$M_k = \sup\{f(x) : x \in I_k\} = f(x_k).$$

By design,

$$\Delta x_k = \frac{b - a}{n} \quad \text{for all } k.$$

We may now compute

$$L(\mathcal{P}) = \sum_{k=1}^{n} m_k \Delta x_k = [f(x_0) + f(x_1) + \cdots + f(x_{n-1})] \cdot \frac{b - a}{n}$$

and

$$U(\mathcal{P}) = \sum_{k=1}^{n} M_k \Delta x_k = [f(x_1) + f(x_2) + \cdots + f(x_n)] \cdot \frac{b - a}{n}.$$

Subtracting, we find that the diameter of $\mathcal{R}(\mathcal{P})$ is

$$U(\mathcal{P}) - L(\mathcal{P}) = [f(x_n) - f(x_0)] \cdot \frac{b - a}{n} = \frac{1}{n}[f(b) - f(a)](b - a).$$

Clearly, we can make this quantity as small as we please by simply choosing n large enough, so we've satisfied our condition for integrability.

The argument for decreasing functions is only slightly different; m_k becomes $f(x_k)$ and M_k becomes $f(x_{k-1})$, so the formula for $U(\mathcal{P}) - L(\mathcal{P})$ becomes $\frac{1}{n}[f(a) - f(b)](b - a)$. ∎

THEOREM 3.2: *Every continuous function on $[a, b]$ is Riemann integrable over $[a, b]$.*

□ *Proof:* The integrability of every function that is continuous on $[a, b]$ is a much deeper result, but we did the hard part in Chapter 2 when we proved Theorem 5.2. Given $\varepsilon > 0$, whenever f is continuous on $[a, b]$ we can find a partition $\mathcal{P}$ such that $M_k - m_k \leq \varepsilon$ for all k. We just choose δ such that every $x, y \in [a, b]$ with $|x - y| < \delta$ satisfy $|f(x) - f(y)| < \varepsilon$, and then choose any partition $\mathcal{P}$ with mesh less than δ. For such a partition we have

$$U(\mathcal{P}) - L(\mathcal{P}) = \sum_{k=1}^{n}(M_k - m_k)\Delta x_k \leq \varepsilon \sum_{k=1}^{n} \Delta x_k = \varepsilon(b - a).$$

Since we can make this last quantity arbitrarily small by choosing ε, we've satisfied the condition for integrability. ∎

Here's another theorem that's easy to prove, thanks to our simplified condition for integrability. It's used primarily for proving things about general integrable functions; with specific functions we can usually recognize integrability over the subinterval in the same way we recognized integrability over the larger interval.

THEOREM 3.3: *Every function that is Riemann integrable over a closed interval $[a, b]$ is Riemann integrable over every closed subinterval of $[a, b]$.*

□ *Proof:* Suppose that f is Riemann integrable over $[a, b]$ and that $[c, d]$ is a given subinterval of $[a, b]$. Since f must be bounded on $[a, b]$, of course it also bounded on $[c, d]$. Given $\varepsilon > 0$, we need to show there is a partition $\mathcal{P}$ of $[c, d]$ with $U(\mathcal{P}) - L(\mathcal{P}) < \varepsilon$. We know that there is a partition $\mathcal{P}'$ of $[a, b]$ with $U(\mathcal{P}') - L(\mathcal{P}') < \varepsilon$ and that $\mathcal{P}'$ has a refinement $\mathcal{P}'' = \{x_k\}_{k=0}^{n}$ that includes c and d. We form $\mathcal{P}$ from $\mathcal{P}''$ by deleting all the points before c or after d. Since all the terms in the sum $\sum_{k=1}^{n}[M_k - m_k]\Delta x_k$ are nonnegative,

$$U(\mathcal{P}) - L(\mathcal{P}) \leq U(\mathcal{P}'') - L(\mathcal{P}'') \leq U(\mathcal{P}') - L(\mathcal{P}') < \varepsilon.$$

That proves the theorem.   ∎

The theorem has a useful companion, often used to evaluate integrals when the formula for $f(x)$ changes at some point in the interval of integration. It also allows us to recognize the integrability of functions that are neither monotonic nor continuous over the entire interval; we split the interval into subintervals where one or the other of those conditions is satisfied.

THEOREM 3.4: *Suppose that $a < b < c$, and that f is Riemann integrable over both $[a, b]$ and $[b, c]$. Then f is Riemann integrable over $[a, c]$ as well, and*

$$\int_a^c f(x)\ dx = \int_a^b f(x)\ dx + \int_b^c f(x)\ dx.$$

☐ *Proof:* Given $\varepsilon > 0$, we can find a partition $\mathcal{P}_1$ of $[a, b]$ with

$$\mathcal{R}(\mathcal{P}_1) \subset \left(\int_a^b f(x)\ dx - \frac{1}{2}\varepsilon, \int_a^b f(x)\ dx + \frac{1}{2}\varepsilon \right)$$

as well as a partition $\mathcal{P}_2$ of $[b, c]$ with

$$\mathcal{R}(\mathcal{P}_2) \subset \left(\int_b^c f(x)\ dx - \frac{1}{2}\varepsilon, \int_b^c f(x)\ dx + \frac{1}{2}\varepsilon \right).$$

We combine $\mathcal{P}_1$ and $\mathcal{P}_2$ to form a partition $\mathcal{P}$ of $[a, c]$, and then observe that every Riemann sum in $\mathcal{R}(\mathcal{P})$ can be split into one Riemann sum in $\mathcal{R}(\mathcal{P}_1)$ plus another in $\mathcal{R}(\mathcal{P}_2)$, so that $\mathcal{R}(\mathcal{P}) \subset (S - \varepsilon, S + \varepsilon)$ with

$$S = \int_a^b f(x)\ dx + \int_b^c f(x)\ dx.$$

Hence the desired conclusion follows from Theorem 2.1.   ∎

Our final theorem in this section is useful for integrating discontinuous functions. When combined with the previous theorem, it shows that changing the value of a function at a finite number of points in an interval affects neither the integrability nor the integral of the function.

THEOREM 3.5: *Suppose that f is Riemann integrable over $[a, b]$, and that g is a function defined on $[a, b]$ that agrees with f on (a, b). Then g is Riemann integrable over $[a, b]$, and*

$$\int_a^b g(x)\ dx = \int_a^b f(x)\ dx.$$

☐ *Proof:* Once again we use Theorem 2.1. We know that f must be bounded on $[a, b]$ and then g must be bounded as well. When $m \leq f(x) \leq M$ for all $x \in [a, b]$, we see that g is bounded by

$$m' = \min\{m, g(a), g(b)\} \quad \text{and} \quad M' = \max\{M, g(a), g(b)\}.$$

Changing f to g in any Riemann sum associated with a partition $\mathcal{P}$ can't change the sum by more than $2(M' - m')$ times the mesh of the partition, and that's the key to the proof. When $\varepsilon > 0$ is given, we find δ with $2(M' - m')\delta \leq \frac{1}{2}\varepsilon$. Then we choose a partition $\mathcal{P}$ with mesh at most δ and having the property that every Riemann sum for f associated with $\mathcal{P}$ is within $\frac{1}{2}\varepsilon$ of $\int_a^b f(x)\, dx$. Every Riemann sum for g associated with $\mathcal{P}$ will then be within ε of $\int_a^b f(x)\, dx$, and that's all we needed to show. ∎

Because of the last theorem, it's common to consider a function as Riemann integrable over an interval in some cases where a finite number of points in the interval are missing from the domain of the function. For example, if $f(x) = |x|$ then $f'(x)$ isn't defined at the origin, yet we can consider f' to be integrable over $[-1, 1]$ because there is an integrable function that agrees with it everywhere except at a single point. But the theorem is no panacea. For example, $1/x$ is continuous everywhere except at the origin, but it isn't integrable over $[0, 1]$; the problem is not the absence of a value at $x = 0$ but the behavior of the function near that point.

EXERCISES

9. Use the intermediate value theorem to show that if f is continuous on $[a, b]$, then there is at least one $c \in (a, b)$ such that

$$\int_a^b f(x)\, dx = f(c)(b - a).$$

   This is the simplest case of the **mean value theorem for integrals.**

10. A more general version of the mean value theorem for integrals states that if f is continuous on $[a, b]$ and if g is positive and Riemann integrable over $[a, b]$, then there is at least one $c \in (a, b)$ such that

$$\int_a^b f(x)g(x)\, dx = f(c)\int_a^b g(x)\, dx.$$

   Of course, the theorem must also apply when g is negative on $[a, b]$ because $-g$ would then be positive. Show by example why it is necessary to assume that g keeps the same sign throughout $[a, b]$.

11. We say that a function f defined on $[a, b]$ is a **step function** if there is a partition $\{x_k\}_{k=1}^{n}$ of $[a, b]$ such that f is constant on each subinterval (x_{k-1}, x_k), $k = 1, \ldots, n$. Explain why step functions are integrable, and find a formula for $\int_a^b f(x)\, dx$ when f is a step function.

12. Prove that every numerical function that is continuous on (a, b) and bounded on $[a, b]$ is Riemann integrable over $[a, b]$. (Most of the terms in the sum for $U(\mathcal{P}) - L(\mathcal{P})$ correspond to a partition of a closed subinterval of (a, b). Use that observation to prove that $\mathcal{P}$ can be chosen to make $U(\mathcal{P}) - L(\mathcal{P})$ arbitrarily small.)

13. Generalize the previous exercise to prove that a bounded function f defined on $[a, b]$ is integrable if it has only finitely many discontinuities.

4 FUNCTIONS DEFINED BY INTEGRALS

So far we've generally been assuming that our interval of integration was fixed, although we have considered the possibility of integrating over subintervals. Now we wish to study new functions whose values are given by integrating a fixed function over a variable subinterval of a fixed interval I. We assume that we have a numerical function f whose domain includes all of I, and that f is Riemann integrable over $[a, b]$ whenever $[a, b] \subset I$. When I is a closed interval we need f to be Riemann integrable over I, but when I isn't closed we need not assume that f is even bounded on I. For example, when f is either continuous or monotonic on I it will be Riemann integrable over every closed subinterval of I. We won't worry further about how we might be able to tell if this condition is satisfied; we'll simply state it as a hypothesis.

The first order of business is to extend the definition of $\int_a^t f(x)\, dx$ to every pair of numbers $a, t \in I$, not just those with $a < t$. Following standard practice, we agree that

$$\int_a^t f(x)\, dx = -\int_t^a f(x)\, dx \quad \text{when } t < a$$

and

$$\int_a^a f(x)\, dx = 0.$$

With that understanding, for all $a, u, v \in I$ we have

$$\int_u^v f(x)\, dx = \int_a^v f(x)\, dx - \int_a^u f(x)\, dx \tag{5.1}$$

regardless of their position in the interval. Proving this relationship is simple but tedious; after we deal with the trivial cases where some of the endpoints are equal there are still six cases to consider. When $a < u < v$, equation (5.1) is a statement about integrals over $[a, u]$, $[u, v,]$, and $[a, v]$, and it agrees with the formula proved in Theorem 3.4. The other cases require an additional step. For example, when $u < v < a$ we need to first rewrite equation (5.1) in terms of integrals over $[u, v]$, $[v, a]$, and $[u, a]$ before appealing to Theorem 3.4.

In view of (5.1), every function defined by integrating f over a variable interval can be treated in terms of $\int_a^t f(x)\ dx$ with a single fixed value of a. We simply use our rules for working with combinations of functions. So all the results we'll ever need can be stated conveniently in the theorem below.

THEOREM 4.1: *Let I be an interval and let f be a numerical function whose domain includes I. Suppose that f is Riemann integrable over every closed subinterval $[a, b]$ of I. Then the formula*

$$F(t) = \int_a^t f(x)\ dx, \quad t \in I$$

defines a continuous numerical function on I. Moreover, F is differentiable at each point in the interior of I where f is continuous, and at those points it satisfies

$$F'(t) = f(t).$$

Before we prove this theorem, let's try to understand its significance. It's almost as important as the formula

$$f'(a) = \lim_{x \to a} \frac{f(x) - f(a)}{x - a}.$$

In particular, when f is continuous on I, it shows us how we can define an **antiderivative** of f on I, that is, a continuous function on I whose derivative is f. It's often useful to be able to refer to a formula for the solution to a problem, and Theorem 4.1 gives the formula used for the problem of finding antiderivatives. It's also a formula that can be used for practical calculations. Just as we can approximate the value of the derivative of f at an arbitrary point a by evaluating a difference quotient $[f(x) - f(a)] / (x - a)$ with x close to a but not exactly equal to it, we can approximate a value of a particular antiderivative by choosing a partition and evaluating a Riemann sum.

☐ *Proof:* We begin the proof by checking the continuity of F at an arbitrary point b in I. The key is to note that for all $t \in I$,

$$F(t) = F(b) + \int_b^t f(x)\, dx.$$

For any $b \in I$ we can find an $r > 0$ such that $I_b = [b - r, b + r] \cap I$ is a closed subinterval of I; it doesn't matter whether b is an interior point or an endpoint of the interval I. Our hypotheses say that f is Riemann integrable over I_b, and so there is a positive number M such that

$$-M \le f(x) \le M \quad \text{for all } x \in I_b.$$

That gives us a simple way to bound the integral of f over subintervals of I_b, and for all $t \in I_b$ we have

$$|F(t) - F(b)| = \left| \int_b^t f(x)\, dx \right| \le M\, |t - b|.$$

To satisfy the definition of continuity of F at b, when we're given $\varepsilon > 0$ we choose δ to be the smaller of r and ε/M.

Next we assume that b is in the interior of I and that f is continuous at b. Then we prove that F is differentiable at b with $F'(b) = f(b)$. This time we can choose $r > 0$ such that $(b - r, b + r) \subset I$, and for all t in that interval except b itself we can write

$$F(t) = F(b) + \int_b^t f(x)\, dx$$

$$= F(b) + (t - b)\left[\frac{1}{t - b} \int_b^t f(x)\, dx \right].$$

So we need to show that the expression in brackets defines a function with a removable discontinuity at $t = b$ and that $f(b)$ is the value we use to remove the discontinuity. We first write

$$\int_b^t f(x)\, dx = \int_b^t f(b)\, dx + \int_b^t [f(x) - f(b)]\, dx$$

$$= (t - b) f(b) + \int_b^t [f(x) - f(b)]\, dx.$$

Since f is continuous at b, for any given $\varepsilon > 0$ there is a $\delta > 0$ such that

$$-\frac{1}{2}\varepsilon < f(x) - f(b) < \frac{1}{2}\varepsilon \quad \text{for every } x \in (b - \delta, b + \delta) \cap I,$$

and it doesn't hurt to take $\delta \le r$. Then for $t \ne b$ in $(b - \delta, b + \delta)$, we have

$$\left| \frac{1}{t-b} \int_b^t f(x) \, dx - f(b) \right| = \left| \frac{1}{t-b} \int_b^t [f(x) - f(b)] \, dx \right| \le \frac{1}{2}\varepsilon < \varepsilon.$$

That's just what we need to complete the proof. ■

While Theorem 4.1 guarantees the differentiability of $\int_a^t f(x) \, dx$ at points where f is continuous, it says nothing about differentiability at other points. Without continuity, differentiability may well fail; for example, consider

$$f(x) = \begin{cases} 1, & x \ge 0 \\ -1, & x < 0. \end{cases}$$

In this case we easily find

$$\int_0^t f(x) \, dx = |t|,$$

and $|t|$ is not differentiable at 0, the discontinuity of f. On the other hand, if we define

$$g(x) = \begin{cases} 1, & x \ne a \\ 2, & x = a, \end{cases}$$

then we find

$$\frac{d}{dt} \int_0^t g(x) \, dx = \frac{d}{dt}(t) = 1 \quad \text{for all } t \in \mathbf{R}.$$

So in this case, differentiating the function defined by the integral removes the discontinuity in the function integrated.

The chain rule provides a useful extension of Theorem 4.1. If $u(t)$ and $v(t)$ are points in the interior of I at which f is continuous, and u and v are differentiable functions, then

$$\frac{d}{dt} \int_{u(t)}^{v(t)} f(x) \, dx = f(v(t)) v'(t) - f(u(t)) u'(t).$$

To prove it, all we do is write

$$\int_{u(t)}^{v(t)} f(x) \, dx = F(v(t)) - F(u(t))$$

and invoke the chain rule.

Theorem 4.1 is one of the basic tools in the study of differential equations, where the basic problem is to find a function whose derivative satisfies a given equation. While we can often recognize antiderivatives by reversing the formulas for derivatives, Theorem 4.1 provides the simplest general formula for defining a function in terms of its derivative.

Some problems in modern applied mathematics appear to involve derivatives of functions that aren't necessarily differentiable. In many of these cases, it turns out that what we really need is the possibility of recovering the function by integrating a nominal derivative, not the possibility of producing a derivative by differentiating the original function. We'll see an example of such a situation when we investigate arc length. A generalized notion of derivative, recently developed, is based on integration instead of differentiation and is often used in such problems. Theorem 4.1 provides the first hints toward such a theory.

EXERCISES

14. For $x \in \mathbf{R}$, define $f(x)$ to be the largest $n \in \mathbf{Z}$ with $n \leq x$, and define $F(t) = \int_0^t f(x)\,dx$. Where is F continuous? Where is it differentiable?

15. What is the largest interval on which $F(t) = \int_0^t \dfrac{dx}{x^2 - a^2}$ defines F as a continuous function?

16. Suppose that f is Riemann integrable over every closed interval $[a, b]$, and F is defined by

$$F(t) = \int_0^t \left[\int_0^s f(x)\,dx \right] ds.$$

Explain why this formula defines $F(t)$ for all $t \in \mathbf{R}$, and why F is differentiable. Where is F' differentiable?

17. For f a numerical function that is continuous on the open interval I and $a \in I$, show that

$$F(t) = \int_a^t (t - x) f(x)\,dx$$

defines a function $F \in C^2(I)$ with

$$F(a) = F'(a) = 0, \quad \text{and} \quad F''(t) = f(t) \text{ for all } t \in I.$$

(Note that Theorem 4.1 doesn't apply directly to $F(t)$ because the function to be integrated depends on t as well as x. Rewrite the formula for $F(t)$ before trying to find $F'(t)$.)

18. For f a numerical function that is continuous on the open interval I and $a \in I$, show that

$$G(t) = \frac{1}{2} \int_a^t (t - x)^2 f(x) \, dx$$

defines a function $G \in C^3(I)$ with

$$G(a) = G'(a) = G''(a) = 0, \quad \text{and} \quad G^{(3)}(t) = f(t).$$

19. The two preceding exercises are special cases of a more general formula. Try to discover it.

5   THE FUNDAMENTAL THEOREM OF CALCULUS

In the previous section we showed how we can use integration to define antiderivatives. Now we'll turn the process around and show how we can use antiderivatives to evaluate integrals. This is considered so important that the theorem below is usually called the **fundamental theorem of calculus**; some authors also consider Theorem 4.1 to be part of the fundamental theorem. Our versions of these theorems are slightly more complicated than those found in basic calculus texts because we want to include the case of discontinuous integrable functions.

THEOREM 5.1:   *Let f be a numerical function that is Riemann integrable over the closed interval $[a, b]$, and let F be a continuous numerical function on $[a, b]$. Suppose that with the possible exception of finitely many points in (a, b), F is differentiable with $F'(x) = f(x)$. Then*

$$\int_a^b f(x) \, dx = F(b) - F(a).$$

If we assume f and F to be continuous on $[a, b]$ with $F'(x) = f(x)$ at all points of (a, b), it's easy to prove Theorem 5.1 by using Theorem 4.1. In that case we can define

$$G(t) = F(t) - \int_a^t f(x) \, dx \quad \text{for } t \in [a, b],$$

making G continuous on $[a, b]$ with $G'(x) = 0$ on (a, b). Then the mean value theorem shows that $G(b) - G(a) = 0$, and therefore

$$F(b) - \int_a^b f(x) \, dx - F(a) = 0.$$

Solving for the integral gives the stated formula. However, we've made no assumption of continuity for f, so a different argument is needed to prove the fundamental theorem in the form we stated.

□ *Proof:* We'll first consider a special case, and then show how to reduce the general case to the one we've considered. When F is differentiable with $F'(x) = f(x)$ at all points in (a, b), we prove that $F(b) - F(a)$ is the value of the integral by showing it is in $\mathcal{R}(\mathcal{P})$ for every partition $\mathcal{P}$, making it the one number in $[L(\mathcal{P}), U(\mathcal{P})]$ for every choice of $\mathcal{P}$. Once $\mathcal{P} = \{x_k\}_{k=1}^n$ has been specified, using the mean value theorem on each subinterval $[x_{k-1}, x_k]$ shows there are sampling points $\{x_k^*\}_{k=1}^n$ with

$$f(x_k^*) \Delta x_k = F(x_k) - F(x_{k-1}) \text{ for } k = 1, \ldots, n.$$

Then

$$\sum_{k=1}^n f(x_k^*) \Delta x_k = \sum_{k=1}^n [F(x_k) - F(x_{k-1})]$$
$$= F(x_n) - F(x_0) = F(b) - F(a),$$

showing that $F(b) - F(a)$ is in $\mathcal{R}(\mathcal{P})$.

For the general case of the fundamental theorem, we list the exceptional points in (a, b) in increasing order; then we can assume

$$a = x_0 < x_1 < x_2 < \cdots < x_n = b$$

with F continuous on each subinterval $[x_{k-1}, x_k]$ and $F'(x) = f(x)$ for all $x \in (x_{k-1}, x_k)$. Then for each k we have

$$\int_{x_{k-1}}^{x_k} f(x) \, dx = F(x_k) - F(x_{k-1}),$$

and so

$$\int_a^b f(x) \, dx = \sum_{k=1}^n \int_{x_{k-1}}^{x_k} f(x) \, dx$$
$$= \sum_{k=1}^n [F(x_k) - F(x_{k-1})] = F(b) - F(a).$$

That takes care of the general case.   ∎

In most calculus classes we're taught to compute $\int_a^b f(x) \, dx$ by simply finding an antiderivative $F(x)$ and then evaluating $F(b) - F(a)$. But there really are hypotheses to be checked, and they're not always satisfied. For example,

$$\int_a^b \frac{dx}{x^2} = -\frac{1}{b} + \frac{1}{a}$$

is only true when $0 \notin [a, b]$ because the function $1/x^2$ isn't integrable over any interval that includes the origin. Besides checking f for integrability over $[a, b]$, at each point where F is not differentiable we need to check F for continuity. Problems can arise in surprising ways. For example,

$$F(x) = \begin{cases} \sqrt{x^2 + 1}, & x \geq 0 \\ -\sqrt{x^2 + 1}, & x < 0 \end{cases}$$

satisfies

$$F'(x) = f(x) = \frac{|x|}{\sqrt{x^2 + 1}} \quad \text{for all } x \neq 0,$$

so F' has a removable discontinuity at $x = 0$ even though the discontinuity of F at 0 is not removable. While $\int_a^b f(x)\, dx$ exists for all $[a, b]$, it's only given by $F(b) - F(a)$ for $0 \notin (a, b]$. Something definitely goes wrong when $a < 0 \leq b$, because in that case $F(b) - F(a)$ is larger than $b - a$ yet all the values of f are between 0 and 1.

Although it's unusual to use Theorem 5.1 to evaluate $\int_a^b f(x)\, dx$ over an interval where f has a discontinuity, there are simple cases where it can be done. For example,

$$f(x) = \begin{cases} 1, & x \geq 0 \\ -1, & x < 0 \end{cases}$$

defines a function with a discontinuity at 0, and $F(x) = |x|$ defines a continuous function with $F'(x) = f(x)$, $x \neq 0$. In this case Theorem 5.1 says that

$$\int_a^b f(x)\, dx = |b| - |a| \quad \text{for all } a, b \in \mathbf{R}.$$

EXERCISES

20. Both $\arctan x$ and $\arctan(-1/x)$ are antiderivatives for $(1 + x^2)^{-1}$. Over which intervals $[a, b]$ can each be used to evaluate

$$\int_a^b \frac{dx}{1 + x^2}?$$

21. Let $F(x) = |x - c| - |x - d|$, where c and d are real constants with $c < d$. Find a function $f(x)$ such that

$$\int_a^b f(x)\, dx = F(b) - F(a) \quad \text{for every closed interval } [a, b].$$

22. The integration-by-parts formula is a consequence of the product rule for derivatives and Theorem 5.1. State appropriate hypotheses for it.

6 TOPICS FOR FURTHER STUDY

There's an "obvious" inequality for integrals that's missing from this chapter: if $b - a > 0$ and f is Riemann integrable on $[a, b]$ with $m < f(x) < M$ for all $x \in [a, b]$, then

$$m(b - a) < \int_a^b f(x)\, dx < M(b - a).$$

Since we haven't assumed that f is continuous, that inequality is remarkably difficult to prove. Note that it implies there are subintervals where the infimum of the values of f is strictly larger than m and others where the supremum of the values of f is strictly smaller than M. An equivalent formulation is a little easier to deal with: if f is nonnegative and Riemann integrable over $[a, b]$ with $\int_a^b f(x)\, dx = 0$, then there is a point in $[a, b]$ where f vanishes. One way to prove it is to show there is a nested sequence of closed intervals $\{[a_n, b_n]\}_{n=1}^\infty$ with

$$\sup\{f(x) : a_n \leq f(x) \leq b_n\} < 1/n;$$

then f must vanish on their intersection.

With a careful application of the tools we've developed, it's possible to prove that if f is Riemann integrable on $[a, b]$ and Φ is uniformly continuous on $\{f(x) : a \leq x \leq b\}$, then $F(x) = \Phi(f(x))$ also defines a Riemann integrable function on $[a, b]$. Proofs appear in many books on real analysis, such as the one by Rudin [5].

Sometimes we integrate functions containing an extra variable that remains fixed while x varies over the interval of integration; this happened in some of the exercises. Leibniz observed that under fairly general hypotheses, we might consider the integral to be a differentiable function of the extra variable, and in those cases Leibniz's formula

$$\frac{d}{dy} \int_a^b F(x, y)\, dx = \int_a^b F_2(x, y)\, dx$$

is often useful.

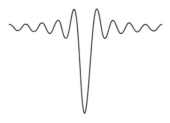

VI
EXPONENTIAL AND
LOGARITHMIC FUNCTIONS

Any serious study of calculus leads to the discovery that many problems can't be solved by working only with functions defined by algebraic formulas. We often need functions that transcend this limitation. Fortunately, calculus provides us with the means to produce **transcendental functions**; in this chapter we'll study the simplest of these.

I EXPONENTS AND LOGARITHMS

Since the laws for working with exponents and logarithms are consequences of the meanings we assign to these terms, we'll begin by recalling how they are developed in algebra. The fundamental purpose of an exponent is to indicate the number of times a factor appears in a product, which accounts for the fundamental law of exponents:

$$(a^m)(a^n) = a^{m+n} \quad \text{for all } m, n \in \mathbf{N}.$$

The fact that mn is the result of adding m to itself n times accounts for the secondary rule:

$$(a^m)^n = a^{mn} \quad \text{for all } m, n \in \mathbf{N}.$$

This exponential notation is used for all sorts of products of mathematical objects, not just multiplication of numbers, so the two laws above appear in a variety of contexts.

In algebra, exponential notation is extended to include exponents that are not natural numbers. When a is a positive real number, there's an algebraic meaning assigned to a^r for every rational number r. (There are also times when fractional powers of negative numbers are used, but we won't consider that case here.) For $a > 0$, we first extend exponential notation to all of $\mathbf{Z}$ by defining

$$a^0 = 1 \quad \text{and} \quad a^{-n} = \frac{1}{a^n}.$$

After checking that the two laws of exponents remain valid for all $m, n \in \mathbf{Z}$, we define rational exponents by calling $a^{m/n}$ the positive solution of the equation $x^n = a^m$ when $m \in \mathbf{Z}$ and $n \in \mathbf{N}$.

However, there's a problem with that process because the equation we're solving to find x depends on m and n separately. To define a^r for $r \in \mathbf{Q}$, the value we assign to it must depend only on the value of r as a rational number, yet there are many different ways to write a single rational number as a fraction. To prove we have a valid definition, it's necessary to show that whenever $m/n = m'/n'$, the equations $x^n = a^m$ and $x^{n'} = a^{m'}$ have the same positive solution. That is, we need to assume that a, x, and y are positive real numbers, that m, m', n, and n' are natural numbers with n and n' positive, and finally that

$$x^n = a^m \text{ and } y^{n'} = a^{m'} \quad \text{with } \frac{m}{n} = \frac{m'}{n'}.$$

We prove $x = y$ as a consequence of those assumptions. We first note that

$$y^{n'n} = \left(y^{n'}\right)^n = \left(a^{m'}\right)^n = a^{m'n}$$

as well as

$$x^{nn'} = \left(x^n\right)^{n'} = \left(a^m\right)^{n'} = a^{mn'}.$$

Since the assumption $m/n = m'/n'$ tells us that $m'n = mn'$, we see that x and y are positive solutions of the same equation—an equation that has just one positive solution. Hence $x = y$, and our definition of rational exponents is a valid one.

The next step is to show that the two basic laws of exponents remain valid for arbitrary rational exponents:

$$(a^r)(a^s) = a^{r+s} \quad \text{for all } r, s \in \mathbf{Q},$$
$$(a^r)^s = a^{rs} \quad \text{for all } r, s \in \mathbf{Q}.$$

Finally, showing that

$$(ab)^r = a^r b^r \quad \text{and} \quad \left(\frac{a}{b}\right)^r = \frac{a^r}{b^r}.$$

whenever a and b are positive real numbers and $r \in \mathbf{Q}$ completes the development of the algebraic theory of exponents.

In algebra, logarithms are introduced in connection with exponents. For a and x positive numbers, it's customary to define

$$\log_a(x) = r \quad \text{if } x = a^r.$$

This definition and the laws of exponents lead to the familiar laws of logarithms. For example, if we assume that

$$\log_a(x) = r \quad \text{and} \quad \log_a(y) = s,$$

then since

$$xy = (a^r)(a^s) = a^{r+s}$$

we have

$$\log_a(xy) = r + s = \log_a(x) + \log_a(y).$$

However, if we regard a^r as defined only when $r \in \mathbf{Q}$, then logarithms become quite problematic because we lack a criterion for recognizing positive real numbers as expressible in the form a^r with r rational. That problem did not block the development of logarithms as a computational tool, and for several hundred years scientists found logarithms by referring to tables and interpolating between the tabulated values. There was an underlying assumption that $\log_a(x)$ is a continuous, monotonic function defined for all $x > 0$. Of course, if $\log_a(x)$ really is a continuous function defined on $(0, \infty)$ then its values can't be limited to the rational numbers; the intermediate value theorem doesn't allow such things. Irrational exponents are needed to complete the theory of logarithms.

EXERCISES

1. When $a < 0$ but a^m has an nth root, this root is sometimes indicated by $a^{m/n}$. Show that with this interpretation, the value of $(-1)^{m/n}$ depends on m and n separately instead of depending only on the value of m/n.

2. Prove that when $a, b > 0$ and $r \in \mathbf{Q}$ we have

$$(ab)^r = a^r b^r.$$

2 ALGEBRAIC LAWS AS DEFINITIONS

As we expanded our notion of exponents from $\mathbf{N}$ to $\mathbf{Q}$, we made sure that the basic laws of exponents and logarithms remained valid. That remains our goal as we expand our notion of exponents to all of $\mathbf{R}$. Instead of worrying immediately about how to define a^x or $\log_a(x)$ for arbitrary x, we'll focus our attention on understanding the properties we expect them to have. Our first step is to define general exponential and logarithmic functions to be otherwise unspecified functions that satisfy appropriate conditions. Then we will develop the properties that such functions must have as a consequence. In the next two sections we'll use the methods of calculus to introduce the most important examples of exponential and logarithmic functions: the **natural exponential function** and the **natural logarithm**. We'll be able to define a^x and $\log_a(x)$ in terms of these two functions.

DEFINITION 2.1:   An **exponential function** is a differentiable function $E : \mathbf{R} \to (0, \infty)$ such that

$$E(x + y) = E(x) E(y) \quad \text{for every } x, y \in \mathbf{R}.$$

A **logarithmic function** is a differentiable function $L : (0, \infty) \to \mathbf{R}$ such that

$$L(xy) = L(x) + L(y) \quad \text{for every } x, y > 0.$$

Since $0 + 0 = 0$ and $1 \cdot 1 = 1$, it's clear that every exponential function satisfies

$$E(0) = 1$$

and every logarithmic function satisfies

$$L(1) = 0.$$

The constant functions E_0 and L_0 defined by $E_0(x) = 1$ and $L_0(x) = 0$ satisfy our definitions of exponential and logarithmic functions. But since they're not of much interest, we'll call these the **trivial** exponential and logarithmic functions. We'll need to establish the existence of nontrivial exponential and logarithmic functions for our definitions have any real use.

We begin by exploring the consequences of our definitions. First we derive some additional equations that must be satisfied by any exponential function E. For any $n \in \mathbf{N}$, the fact that adding any x to itself n times produces nx leads easily to the identity

$$E(nx) = [E(x)]^n. \tag{6.1}$$

A formal proof can be worked out using mathematical induction. Next, since

$$1 = E(0) = E(-nx + nx) = E(-nx)E(nx),$$

we note that equation (6.1) is valid for all $n \in \mathbf{Z}$ as well as $n \in \mathbf{N}$. Thus for all m/n with $m \in \mathbf{Z}$ and $n \in \mathbf{N}$,

$$[E(m/n)]^n = E(m) = [E(1)]^m,$$

so our definition of rational exponents shows that

$$E(r) = [E(1)]^r \quad \text{for every } r \in \mathbf{Q}.$$

That's why we called a function E that satisfies our definition an exponential function. The number $E(1)$ is called its **base**.

Now let L be an arbitrary logarithmic function. For any $x > 0$, we know that

$$L(x^2) = L(x \cdot x) = L(x) + L(x) = 2L(x),$$

and, more generally, that

$$L(x^n) = nL(x) \quad \text{for every } n \in \mathbf{N}.$$

Replacing x by $x^{1/n}$ shows that

$$L(x) = nL\left(x^{1/n}\right),$$

and then replacing x by x^m with $m \in \mathbf{N}$ shows that

$$L\left(x^{m/n}\right) = \frac{1}{n}L(x^m) = \frac{m}{n}L(x).$$

Since

$$0 = L(1) = L\left(x^0\right) = L\left(x^{-m/n}x^{m/n}\right) = L\left(x^{-m/n}\right) + L\left(x^{m/n}\right),$$

we see that

$$L\left(x^r\right) = rL\left(x\right) \quad \text{for every } x > 0 \text{ and every } r \in \mathbf{Q}.$$

When L is a nontrivial logarithmic function, its set of values must be all of $\mathbf{R}$. Given any $y \in \mathbf{R}$, once we pick $a > 0$ with $L\left(a\right) \neq 0$ we can find a positive integer n such that $n\left|L\left(a\right)\right| > |y|$. Then y must be between $L\left(a^{-n}\right)$ and $L\left(a^n\right)$. Since L is differentiable, it is continuous. So the intermediate value theorem guarantees that there is an x between a^{-n} and a^n with $L\left(x\right) = y$.

In particular, whenever L is a nontrivial logarithmic function there must be a positive number b with $L\left(b\right) = 1$, and we can then write

$$L\left(b^r\right) = r \quad \text{for every } r \in \mathbf{Q}.$$

That justifies our calling L a logarithmic function; the number b is called its **base**.

Using Definition 2.1, it's easy to derive a formula for derivatives of arbitrary exponential functions. First, note that if E is any exponential function and a is any constant, then for all $x \in \mathbf{R}$ we have

$$E\left(x\right) = E\left(a\right)E\left(x - a\right).$$

Consequently, with the help of the chain rule we find that

$$E'\left(x\right) = E\left(a\right)E'\left(x - a\right)\frac{d}{dx}\left(x - a\right) = E\left(a\right)E'\left(x - a\right).$$

Then setting $x = a$ shows that

$$E'\left(a\right) = E\left(a\right)E'\left(0\right).$$

Since a was an arbitrary real number, we must have

$$E'\left(x\right) = E'\left(0\right)E\left(x\right) \text{ for all } x \in \mathbf{R}. \tag{6.2}$$

Now let's find an equation satisfied by the derivative of any logarithmic function L. According to Definition 2.1, for any constant $a > 0$ and for all $x > 0$ we can write

$$L\left(x\right) = L\left(\frac{x}{a}\right) + L\left(a\right).$$

This time we find that

$$L'(x) = L'\left(\frac{x}{a}\right)\frac{d}{dx}\left(\frac{x}{a}\right) = \frac{1}{a}L'\left(\frac{x}{a}\right).$$

Again setting $x = a$, we obtain

$$L'(a) = \frac{1}{a}L'(1).$$

Since a can be any positive real number, we conclude that

$$L'(x) = \frac{L'(1)}{x} \quad \text{for all } x > 0. \tag{6.3}$$

Next we show that exponential and logarithmic functions are characterized by equations (6.2) and (6.3) and their values at a single point. That's the content of the theorem below.

THEOREM 2.1: *A differentiable function $E : \mathbf{R} \to (0, \infty)$ with $E(0)$ $= 1$ is an exponential function if and only if $E'(x)/E(x)$ is constant.*
    A differentiable function $L : (0, \infty) \to \mathbf{R}$ with $L(1) = 0$ is a logarithmic function if and only if $xL'(x)$ is constant.

    □   *Proof:*   We've already seen that $E'(x)/E(x)$ is the constant $E'(0)$ whenever E is an exponential function. Now let's assume that E is a differentiable function with $E(x) > 0$ for all $x \in \mathbf{R}$ and that a is an arbitrary constant. Then we define $f : \mathbf{R} \to (0, \infty)$ by the formula

$$f(x) = \frac{E(x+a)}{E(x)}.$$

According to the quotient rule and the chain rule,

$$f'(x) = \frac{E(x)E'(x+a) - E(x+a)E'(x)}{E(x)^2}$$

$$= \frac{E(x+a)}{E(x)}\left[\frac{E'(x+a)}{E(x+a)} - \frac{E'(x)}{E(x)}\right].$$

Thus if $E'(x)/E(x)$ is constant, then $f'(x) = 0$ for all x. The mean value theorem then guarantees that f is constant, and so for all $x \in \mathbf{R}$ we have

$$\frac{E(x+a)}{E(x)} = f(x) = f(0) = \frac{E(a)}{E(0)}.$$

If we further assume that $E(0) = 1$, we have

$$E(x + a) = E(x) E(a) \quad \text{for all } x, a \in \mathbf{R},$$

proving that E is an exponential function.

Before we stated the theorem, we saw that $xL'(x)$ is constant for every logarithmic function L. On the other hand, if we simply assume that L is a differentiable function on $(0, \infty)$, then for any $a > 0$ we can define another differentiable function $g : (0, \infty) \to \mathbf{R}$ by the formula

$$g(x) = L(ax) - L(x).$$

By the chain rule,

$$g'(x) = aL'(ax) - L'(x),$$

and so multiplication by x yields

$$xg'(x) = axL'(ax) - xL'(x).$$

Consequently, if $xL'(x)$ is constant, then $g'(x) = 0$ for all $x > 0$. So $g(x)$ must be constant as well. In that case, for all $x > 0$ we have

$$L(ax) - L(x) = g(x) = g(1) = L(a) - L(1).$$

With the assumption that $L(1) = 0$, we then find that

$$L(ax) = L(a) + L(x) \quad \text{for all } a, x > 0,$$

proving that L is a logarithmic function. ■

EXERCISES

3. Prove that the product of two exponential functions is also an exponential function.

4. Suppose that E is a nontrivial exponential function. Find all the differentiable functions $f : \mathbf{R} \to \mathbf{R}$ such that the composite function $E \circ f$ is also a nontrivial exponential function.

5. Suppose that L is a nontrivial logarithmic function. Find all the differentiable functions $f : \mathbf{R} \to \mathbf{R}$ such that the composite function $f \circ L$ is also a nontrivial logarithmic function.

6. Prove that if E is a nontrivial exponential function, then it has an inverse and its inverse is a logarithmic function.

3   THE NATURAL LOGARITHM

In the last section we talked about general exponential and logarithmic functions, but the only ones we've seen so far are the trivial ones. Now we begin to remedy that situation. There's a remarkably simple way to produce a logarithmic function, a way that seems to have little to do with either exponents or logarithms.

DEFINITION 3.1:   The **natural logarithm** is the function log : $(0, \infty) \to \mathbf{R}$ defined by the formula

$$\log(x) = \int_1^x t^{-1} dt.$$

Since $f(t) = t^{-1}$ defines a continuous function on $(0, \infty)$, Theorem 4.1 in Chapter 5 shows that the natural logarithm is a differentiable function, with its derivative given by

$$\log'(x) = x^{-1} \quad \text{for all } x > 0.$$

It also satisfies

$$\log(1) = \int_1^1 t^{-1} dt = 0,$$

so Theorem 2.1 shows that the natural logarithm really is a logarithmic function. It's certainly not the trivial function, since its derivative is never zero.

The base of the natural logarithm is ordinarily called e and is defined by the equation

$$\log(e) = 1.$$

To estimate e, we examine values of the natural logarithm. The integral defining $\log(2)$ shows us that $\frac{1}{2} \leq \log(2) \leq 1$, so therefore $1 \leq \log(4) \leq 2$ and it follows that

$$2 \leq e \leq 4.$$

Other characterizations of e have been used to prove that it is an irrational number, and its decimal representation has been calculated to an enormous number of places.

It isn't terribly difficult to compute the value of $\log x$ quite accurately when x is a given number. Riemann sums provide a natural way to do this, and with the right partition and choice of sampling points they're easy to calculate. First let's suppose that $x > 1$. Instead of using subintervals of equal length, we partition $[1, x]$ into n subintervals by defining

$$t_k = x^{k/n} \quad \text{for } k = 0, \ldots, n.$$

For this partition, the length of the kth subinterval is

$$\Delta t_k = x^{k/n} - x^{(k-1)/n} = x^{(k-1)/n}\left[x^{1/n} - 1\right],$$

and so the mesh of this partition is then

$$x^{1-1/n}\left[x^{1/n} - 1\right].$$

Since $x^{1/n} \to 1$ as $n \to \infty$, we see that we can make the mesh of this partition arbitrarily small by choosing a large enough value for n. We choose our sampling points as

$$t_k^* = t_{k-1} = x^{(k-1)/n},$$

and then the corresponding Riemann sum is

$$\sum_{k=1}^{n} \frac{1}{t_k^*}\Delta t_k = \sum_{k=1}^{n} \frac{x^{(k-1)/n}\left[x^{1/n} - 1\right]}{x^{(k-1)/n}} = n\left[x^{1/n} - 1\right].$$

If $0 < x < 1$ we modify the process slightly, but it isn't really very different. In this case

$$\log(x) = \int_x^1 -t^{-1}dt,$$

so our interval of integration is $[x, 1]$ instead. We define

$$t_k = x^{1-k/n} \text{ for } k = 0, \ldots, n$$

so that

$$\Delta t_k = x^{1-k/n} - x^{1-(k-1)/n} = x^{1-k/n}\left[1 - x^{1/n}\right].$$

In this case we choose

$$t_k^* = t_k = x^{1-k/n}$$

to obtain the same approximation that we found before.

Thus for all $x > 0$, we obtain the formula

$$\log (x) = \lim_{n \to \infty} n \left[x^{1/n} - 1 \right].$$

There's even an efficient way to calculate this limit by using repeated square roots; it amounts to considering only powers of 2 for n. If we define

$$L_k (x) = 2^k \left[x^{1/2^k} - 1 \right],$$

then $x^{1/2^k} = 1 + 2^{-k} L_k (x)$ and $x^{1/2^{k+1}}$ is its square root. Consequently,

$$L_{k+1} (x) = 2^{k+1} \left[\sqrt{1 + 2^{-k} L_k (x)} - 1 \right],$$

so successive approximations can be calculated quite easily. The calculations begin with

$$L_1 (x) = 2 \left[\sqrt{x} - 1 \right].$$

A better way to do the calculations is to rewrite

$$L_{k+1} (x) = 2^{k+1} \left[\frac{1 + 2^{-k} L_k (x) - 1}{\sqrt{1 + 2^{-k} L_k (x)} + 1} \right] = \frac{2 L_k (x)}{1 + \sqrt{1 + 2^{-k} L_k (x)}}.$$

As $k \to \infty$, $\sqrt{1 + 2^{-k} L_k (x)}$ becomes increasingly difficult to distinguish from 1, and eventually both digital computers and calculators will treat them as exactly equal. At that stage, the first formula returns $L_{k+1} (x) = 0$, but the second returns $L_{k+1} (x) = L_k (x)$. Unless x is either extremely large or extremely close to 0, this stage can be reached in a reasonable number of steps.

EXERCISES

7. Prove that if $1 \le x < y$, then $\log y - \log x < y - x$. Does the last inequality remain true for $0 < x < y \le 1$ or for $0 < x < 1 < y$?

8. By writing $\log (|f (x)|)$ as $\frac{1}{2} \log \left([f (x)]^2 \right)$, show that when $f (x)$ is a nonzero differentiable function we always have

$$\frac{d}{dx} \log (|f (x)|) = \frac{f' (x)}{f (x)}.$$

9. Show that for $x > 1$ the sequence $\{L_k (x)\}_{k=1}^{\infty}$ is a decreasing sequence with positive terms and that for $0 < x < 1$ it is a decreasing sequence with negative terms.

10. Show that $e > 2.5$ by finding an integer k such that $L_k (2.5) < 1$.

4 THE NATURAL EXPONENTIAL FUNCTION

While we've characterized an exponential function as a differentiable function $E : \mathbf{R} \to (0, \infty)$ such that $E(0) = 1$ and $E'(x) / E(x)$ is constant, we haven't yet produced a nontrivial one. There are quite a few processes that produce them, but with most of these methods it's very difficult to prove that the function they produce really is an exponential function. There is one notable exception, and that's the route we'll follow. We'll show that the inverse of the natural logarithm is an exponential function, commonly known as the **natural exponential function**. But defining a function as the inverse of a function whose values we can only approximate isn't very satisfying. So in the next section we'll also develop a method for expressing the value of the natural exponential function directly as the limit of a sequence of approximations, with each approximation involving only a finite number of arithmetic operations.

DEFINITION 4.1: The inverse of the natural logarithm is called the **natural exponential function**. We denote the natural exponential function by exp, so that

$$y = \log(x) \text{ if and only if } x = \exp(y).$$

Since the natural logarithm has domain $(0, \infty)$ with no critical points and its set of values is all of $\mathbf{R}$, the inverse function theorem tells us that $\exp : \mathbf{R} \to (0, \infty)$ with

$$\exp'(y) = \frac{1}{\log'(x)} = x = \exp(y)$$

when $y = \log(x)$. Hence $\exp'(y) / \exp(y)$ is indeed constant. We also know $\exp(0) = 1$ because $\log(1) = 0$, so the inverse of the natural logarithm satisfies our definition of exponential function.

The natural logarithm and the natural exponential function give us an easy way to interpret irrational exponents. Since

$$\log(a^r) = r \log(a) \quad \text{when } a > 0 \text{ and } r \in \mathbf{Q},$$

for all such a and r we have

$$a^r = \exp[r \log(a)].$$

We simply agree that

$$a^x = \exp[x \log(a)] \quad \text{for } a > 0 \text{ and } x \in \mathbf{R}.$$

This agrees with the algebraic definition of exponents when x is rational, and since

$$\frac{d}{dx}\left(a^{x}\right) = \frac{d}{dx}\left\{\exp\left[x\log\left(a\right)\right]\right\}$$

$$= \exp'\left[x\log\left(a\right)\right]\frac{d}{dx}\left[x\log\left(a\right)\right]$$

$$= \exp\left[x\log\left(a\right)\right]\log\left(a\right)$$

$$= a^{x}\log\left(a\right),$$

we see that it is indeed an exponential function. The usual laws of exponents are then valid with arbitrary real exponents:

$$\left(a^{x}\right)\left(a^{y}\right) = \exp\left[x\log\left(a\right)\right]\exp\left[y\log\left(a\right)\right]$$

$$= \exp\left[x\log\left(a\right) + y\log\left(a\right)\right]$$

$$= \exp\left[(x + y)\log\left(a\right)\right] = a^{x+y}$$

as well as

$$\left(a^{x}\right)^{y} = \exp\left[y\log\left(a^{x}\right)\right] = \exp\left[xy\log\left(a\right)\right] = a^{xy}.$$

This extension of the definition of a^{x} to the case of arbitrary exponents resolves the difficulty with the definition of $\log_{a}\left(x\right)$. We can again define

$$\log_{a}\left(x\right) = y \quad \text{if and only if } x = a^{y},$$

and since

$$\log\left(a^{y}\right) = y\log\left(a\right)$$

we see that

$$\log_{a}\left(x\right) = \frac{\log\left(x\right)}{\log\left(a\right)}.$$

That makes

$$\frac{d}{dx}\left[\log_{a}\left(x\right)\right] = \frac{1}{\log\left(a\right)}\frac{d}{dx}\left[\log\left(x\right)\right] = \frac{1}{x\log\left(a\right)}.$$

Consequently, we see that this version of $\log_{a}$ satisfies our definition of logarithmic function.

One important application of our definition of a^{x} is the case $a = e$, the base of the natural logarithm. Since e is defined by the condition

$$\log\left(e\right) = 1,$$

we see that

$$e^x = \exp\left[x \log\left(e\right)\right] = \exp\left(x\right).$$

Consequently, the natural exponential function is often expressed in exponential notation rather than functional notation.

EXERCISES

11. Why didn't we use the equation $E'\left(x\right) = E\left(x\right) E'\left(0\right)$ to define $\exp\left(x\right)$ directly, in much the same way we defined $\log\left(x\right)$?

12. Prove that $\left(ab\right)^x = \left(a^x\right)\left(b^x\right)$ for all $a, b > 0$ and all $x \in \mathbf{R}$.

13. Show that if E is any exponential function, then $\log\left[E\left(x\right)\right]$ has a constant derivative. Then explain why every exponential function can be expressed in the form $E\left(x\right) = \exp\left(ax\right)$ for some constant a.

14. Suppose that f and g are differentiable functions, with f positive. Show that

$$\frac{d}{dx}\left\{[f\left(x\right)]^{g\left(x\right)}\right\} = [f\left(x\right)]^{g\left(x\right)}\left\{f'\left(x\right)\frac{g\left(x\right)}{f\left(x\right)} + g'\left(x\right)\log\left[f\left(x\right)\right]\right\}.$$

15. The hyperbolic sine and hyperbolic cosine are defined by the formulas

$$\sinh t = \tfrac{1}{2}\left(e^t - e^{-t}\right) \quad \text{and} \quad \cosh t = \tfrac{1}{2}\left(e^t + e^{-t}\right).$$

Show that each point on the right half of the hyperbola $x^2 - y^2 = 1$ has the form $\left(\cosh t, \sinh t\right)$ for exactly one $t \in \mathbf{R}$. Then show that

$$\frac{d}{dt}\left(\sinh t\right) = \cosh t \quad \text{and} \quad \frac{d}{dt}\left(\cosh t\right) = \sinh t.$$

5 AN IMPORTANT LIMIT

We saw that the natural logarithm can be computed from the formula

$$\log\left(x\right) = \lim_{n \to \infty} n\left[x^{1/n} - 1\right]$$

for any $x > 0$. Consequently, for any $x \in \mathbf{R}$ we have

$$\lim_{n \to \infty} n\left[e^{x/n} - 1\right] = \log\left(e^x\right) = x.$$

This says that as n increases, the difference between $e^{x/n}$ and $1 + \frac{1}{n}x$ becomes negligible in comparison to $1/n$. Since

$$e^x = \left(e^{x/n}\right)^n,$$

can we say that

$$e^x = \lim_{n \to \infty} \left(1 + \frac{1}{n}x\right)^n ?$$

The answer is an emphatic yes, and there are many ways to prove it. We'll prove a more general result:

$$\lim_{t \to 0} (1 + at)^{1/t} = e^a \quad \text{for all } a \in \mathbf{R}.$$

We begin by noting that

$$f(t) = \log(1 + at)$$

defines a function that is differentiable at $t = 0$, and

$$f'(t) = \frac{1}{1 + at} \cdot \frac{d}{dt}(at) = \frac{a}{1 + at} \quad \text{for } 1 + at > 0.$$

Since $f'(0) = a$, the definition of the derivative tells us that

$$a = \lim_{t \to 0} \frac{f(t) - f(0)}{t - 0} = \lim_{t \to 0} \frac{\log(1 + at)}{t}.$$

We know that the natural exponential function is continuous, so

$$\exp(a) = \lim_{t \to 0} \exp\left[\frac{\log(1 + at)}{t}\right] = \lim_{t \to 0} (1 + at)^{1/t}.$$

The limit of the sequence $\left\{\left(1 + \frac{1}{n}x\right)^n\right\}_{n=1}^{\infty}$ is sometimes used to define e^x. It really is possible to prove that this sequence converges for each x and that its limit defines an exponential function. But it's a lengthy process, and we won't begin it; one definition of the natural exponential function is enough. However, we will examine the sequence further and derive an estimate for the error in approximating e^x by the polynomial $\left(1 + \frac{1}{n}x\right)^n$.

Since $e^x = \left(e^{x/n}\right)^n$, the mean value theorem tells us that

$$e^x - \left(1 + \frac{1}{n}x\right)^n = n\xi^{n-1}\left[e^{x/n} - \left(1 + \frac{1}{n}x\right)\right] \tag{6.4}$$

for some ξ between $\left(1 + \frac{1}{n}x\right)$ and $e^{x/n}$. Since

$$\exp'(0) = \exp(0) = 1,$$

we recognize that $1 + \frac{1}{n}x$ is just the linearization of exp near 0, evaluated at x/n instead of x. That suggests using Theorem 4.1 in Chapter 4 to estimate the last factor in equation (6.4). We can write

$$e^{x/n} = 1 + \frac{1}{n}x + \frac{1}{2}\left(\frac{x}{n}\right)^2 \exp(\eta)$$

for some real number η between 0 and x/n. Substituting this into (6.4) gives us the representation

$$e^x - \left(1 + \frac{1}{n}x\right)^n = \frac{1}{2n}x^2\xi^{n-1}\exp(\eta).$$

Since $\exp(\eta)$ is between 1 and $e^{x/n}$, it's reasonable to estimate both ξ and $\exp(\eta)$ by $e^{x/n}$ as $n \to \infty$. We then obtain

$$e^x - \left(1 + \frac{1}{n}x\right)^n \approx \frac{1}{2n}x^2 e^x \text{ as } n \to \infty. \qquad (6.5)$$

In other words, the relative error in this approximation to e^x is about $x^2/2n$ for large values of n. Unless x is quite close to 0, extremely large values of n will be needed to produce high levels of accuracy.

EXERCISES

16. In calculating compound interest, interest rates are stated in terms of a nominal annual interest rate r. If the year is divided into n equal interest periods, the interest rate per interest period is r/n. Calculate the total interest paid over a year on a unit investment, and show that it approaches $e^r - 1$ as the number of interest periods increases without bound. This limiting case is usually called **continuous compounding**.

17. Prove that $\lim_{x\to\infty} x^n/e^x = 0$ for each $n \in \mathbf{N}$.

18. To ten significant figures, $e = 2.718281828$. Find $\left(1 + \frac{1}{n}\right)^n$ for $n - 10$ and $n = 100$, and compare the error to the error estimate for e^1 given by (6.5).

19. Use a quadratic approximation to the natural exponential function to show that for all $x > 0$ and each positive integer n,

$$n\left[x^{1/n} - 1\right] = \log(x) + \frac{1}{2n}\xi \left[\log(x)\right]^2$$

for some ξ between 1 and $x^{1/n}$. Note $x^{1/n} = \exp\left[\frac{1}{n}\log(x)\right]$.

VII

CURVES AND ARC LENGTH

In the previous chapter we used calculus methods to produce functions of a decidedly algebraic nature. Here we'll see that the methods of calculus can also be used to develop some of the basic functions that we use in geometry and trigonometry to find distances.

I   THE CONCEPT OF ARC LENGTH

Clearly the phrase *arc length of a curve* refers to some sort of one-dimensional measurement. But just what is a curve, and what do we mean by its arc length? We'll try to answer these questions by looking at some simple geometric ideas.

If our primary interest in curves is in determining the length of a path that follows one, then we see that it's entirely reasonable for curves to be straight or to have corners, and the mathematical definition of a curve does indeed allow for those possibilities. We normally think of a curve as linking its endpoints in a continuous path, with the points on the curve arranged in a definite order. A natural way to specify this order is to represent the curve in the form

$$x = f(t), y = g(t) \quad \text{for } a \leq t \leq b,$$

with f and g continuous numerical functions on $[a, b]$. This is called a **parametric representation** of the curve; the auxiliary variable t is called a **parameter**. Given such a representation, we can write

$$P(t) = (f(t), g(t))$$

for a generic point on the curve. That orders the points: $P(t_1)$ precedes $P(t_2)$ if t_1 and t_2 are in $[a, b]$ with $t_1 < t_2$. The first point in the curve is $P(a)$, and the last point is $P(b)$.

Specifying a curve in this manner gives us a standard way to construct simple approximate paths. Suppose we partition $[a, b]$ as we did in defining Riemann sums:

$$a = t_0 < t_1 < t_2 < \cdots < t_n = b.$$

Calling P_k the point $P(t_k)$, we can approximate the curve by the broken-line path

$$\overline{P_0 P_1}, \ \overline{P_1 P_2}, \ldots, \overline{P_{n-1} P_n}$$

formed by connecting the successively numbered points with line segments. For any given any $\varepsilon > 0$, Theorem 5.2 in Chapter 2 shows that it's possible to choose the partition in such a way that neither $f(t)$ nor $g(t)$ will change by more than ε over any of the subintervals. So we can choose a broken-line path that stays arbitrarily close to the curve.

When we subdivide a curve into segments, the arc length of the curve should be the same as the sum of the arc lengths of the segments, and the arc length of each segment must be at least the distance between its endpoints. So if we call $|PQ|$ the distance between the points P and Q, the arc length of the curve described above must be at least

$$|P_0 P_1| + |P_1 P_2| + \cdots + |P_{n-1} P_n|$$

for any partition of $[a, b]$, and the sum is exactly the distance along the broken-line path linking $P_0, P_1, \ldots, P_n$ in that order. Since we can approximate general curves by broken-line paths, it's natural to use these sums to define arc length.

It doesn't take long to make an unpleasant discovery—we can't always define arc length. For example, let's consider a simple rectangular spiral that alternates horizontal and vertical segments, with the length of the nth segment given by

$$s_n = \sqrt{n} - \sqrt{n-1}.$$

The length of the first segment is 1, and the lengths decrease monotonically to 0 as n increases. Drawing the whole curve is problematic, but Figure 7.1 shows a sketch of the first ten segments. It's easy to calculate that the sum of the lengths of the first n segments is exactly $\sqrt{n}$, so the entire path can't have a finite length, even though it spirals in to a single point without ever crossing itself.

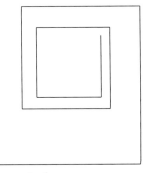

Figure 7.1   A rectangular spiral.

Now we do the sort of thing mathematicians always do in these situations. Along with our formal definition of arc length, we define a restricted class of curves excluding all those curves that can't have an arc length.

DEFINITION 1.1: A curve is said to be **rectifiable** if there is a fixed number that is never exceeded by the sum of the distances between successive points on the curve, no matter how many points are considered. The **arc length** of a rectifiable curve $\mathcal{C}$ is

$$L = \sup \left\{ \sum_{k=1}^{n} |P_{k-1}P_k| : P_0, P_1, \ldots, P_n \text{ successive points on } \mathcal{C} \right\}.$$

When we say that $P_0, P_1, \ldots, P_n$ are successive points on $\mathcal{C}$, that means the points are on $\mathcal{C}$ and are encountered in that order; of course there will always be additional points between them. For the purposes of our definition, it's vital that we keep the points in the same order as they appear on $\mathcal{C}$. It is quite possible for the same set of points to be retraced several times by a rectifiable curve. With our definition, each retracing adds to the arc length.

Let's see how the usual notion of the circumference of a circle matches up with the definition of arc length we've just given. Classically, the circumference of a circle can be approximated from below by the perimeter of inscribed polygons and from above by the perimeter of circumscribed ones. For example, Archimedes used that approach to determine π fairly accurately. We would like to show that circles are rectifiable curves and that the circumference of a circle is the same as its arc length.

We begin by examining sums of distances between successive points on a single tracing of a circle. Since we're interested in making such sums as large as possible, we might as well assume that the last point selected is the same as the first, so that we've gone completely around the circle. In that case, $P_0 = P_n$, the line segments $\overline{P_0 P_1}$, $\overline{P_1 P_2}, \ldots, \overline{P_{n-1} P_n}$ are the sides of an inscribed polygon, and the sum of their lengths is its perimeter. So we can prove the circle is a rectifiable curve if we can find an upper bound for the perimeters of all inscribed polygons.

In fact, the perimeter of a polygon inscribed in a circle is never more than the perimeter of any circumscribed polygon. Although that seems obvious, it's actually fairly subtle, because it's hard to recognize the reasons it seems obvious to us. Fortunately, it's easy to compare each side of an inscribed polygon to part of the perimeter of an circumscribed one. Let's say that P_{k-1} and P_k are successive vertices on the inscribed polygon, and O is the center of the circle. We draw rays from O through P_{k-1} and P_k and extend them until they meet the circumscribed polygon; call those points Q_{k-1} and Q_k. Because the angles $OP_k P_{k-1}$ and $OP_{k-1} P_k$ are equal and therefore acute, it follows that

$$|P_{k-1} P_k| < |Q_{k-1} Q_k| \quad \text{unless } Q_{k-1} = P_{k-1} \text{ and } Q_k = P_k,$$

and $|Q_{k-1} Q_k|$ is never more than the length of any path from Q_{k-1} to Q_k.

That proves that any circle is indeed a rectifiable curve; its arc length is the supremum of the set of perimeters of inscribed polygons and is never more than the perimeter of any circumscribed polygon. What about approximating the arc length by the perimeters of circumscribed polygons? To see how that works, let's suppose that our circle has radius r and we've inscribed a regular polygon with n sides and perimeter p. By the Pythagorean theorem, the distance from the center of the circle to each side is $\sqrt{r^2 - s^2}$, where s is half the length of a single side. So expanding the inscribed polygon by a factor of $r/\sqrt{r^2 - s^2}$ produces a circumscribed polygon. That expansion increases the perimeter by the same factor, and the factor approaches 1 as the number of sides in the regular polygon increases. Consequently, as the number of sides increases, the perimeters of both inscribed and circumscribed regular polygons do indeed approach the arc length of the circle.

Now let's return to the case of general rectifiable curves and make a few observations about the sums we use to approximate their arc length. Given a collection of successive points on a curve, we can always expand it by fitting an additional point into its proper place. If that happens to be at either end of the list, it simply adds in another distance. Otherwise, one of the distances gets replaced by the sum of two, and that sum is at least as big as the distance it replaces. Consequently, expanding the list of successive points on a curve can only increase the sum of the distances between them. That's a good thing to remember, as it plays a crucial role in the theorem below. Note that the theorem requires a careful interpretation if the curve passes through any point more than once, but the interpretation is a natural one for curves given parametrically.

THEOREM 1.1:   *Suppose that P is a point on a curve C, and that C_1 and C_2 are the curves formed by the points on C before and after P. Then C is rectifiable if and only if both C_1 and C_2 are rectifiable, and in that case the arc length of C is the sum of the arc lengths of C_1 and C_2.*

□   *Proof:*   First assume that C is rectifiable, and call L its arc length. Suppose we have successive points $P_0, P_1, \ldots, P_n$ on C_1 and $Q_0, Q_1, \ldots, Q_m$ on C_2; we need upper bounds for $\sum_{k=1}^{n} |P_{k-1}P_k|$ and $\sum_{k=1}^{m} |Q_{k-1}Q_k|$ to prove that C_1 and C_2 are rectifiable. Since adding P to either collection of successive points can only increase the corresponding sum, the only case we really need to consider is when $P_n = P = Q_0$. In that case $P_0, P_1, \ldots, P_n, Q_1, Q_2, \ldots, Q_m$ are successive points on C, and since $P_n = Q_0$ we see that

$$\sum_{k=1}^{n} |P_{k-1}P_k| + \sum_{k=1}^{m} |Q_{k-1}Q_k| \leq L.$$

Rewriting this inequality in the equivalent form

$$\sum_{k=0}^{n} |P_{k-1}P_k| \leq L - \sum_{k=0}^{m} |Q_{k-1}Q_k|$$

shows that C_1 is rectifiable, with arc length

$$L_1 \leq L - \sum_{k=0}^{m} |Q_{k-1}Q_k|.$$

Rewriting this as

$$\sum_{k=0}^{m} |Q_{k-1}Q_k| \leq L - L_1$$

shows that C_2 is also rectifiable, with arc length

$$L_2 \leq L - L_1.$$

That proves half of the theorem.

For the other half, we assume that both C_1 and C_2 are rectifiable with arc lengths L_1 and L_2. Then for $P_0, P_1, \ldots, P_n$ successive points on C, we must have

$$\sum_{k=0}^{n} |P_{k-1}P_k| \leq L_1 + L_2.$$

When P is one of the points in $P_0, P_1, \ldots, P_n$ we can split the sum into two parts, with one bounded by L_1 and the other by L_2. When P isn't one of them, we simply insert P in its proper place and the increased sum has the desired bound. Thus C is rectifiable, with arc length $L \leq L_1 + L_2$. Combining the two inequalities we've proved for L, L_1, and L_2 shows that $L = L_1 + L_2$, and that completes the proof. ∎

When we start cutting a rectifiable curve into pieces, it's convenient to have a function that describes the result. For a rectifiable curve C defined by a pair of continuous numerical functions on $[a, b]$, we can define an **arc-length function** λ on $[a, b]$ by calling $\lambda(t)$ the arc length of the part of C between $P(a)$ and $P(t)$. Clearly λ is an increasing numerical function on $[a, b]$, and by the last theorem $|\lambda(s) - \lambda(t)|$ is the arc length of the part of C between $P(s)$ and $P(t)$. Such an arc length function is always continuous on $[a, b]$; that's important enough that we'll state it as a formal theorem.

THEOREM 1.2: *Let f and g be continuous numerical functions on $[a, b]$, and let C be the curve defined by*

$$P(t) = (f(t), g(t)) \quad \text{for } a \leq t \leq b.$$

If C is rectifiable, then the length of the portion of C between $P(a)$ and $P(t)$ defines a continuous numerical function of t on $[a, b]$.

☐ *Proof:* Calling the arc-length function λ, we'll show that for any given $\varepsilon > 0$ there is a $\delta > 0$ such that

$$|\lambda(s) - \lambda(t)| < \varepsilon \quad \text{for all } s, t \in [a, b] \text{ with } |s - t| < \delta.$$

Since λ is an increasing function, we can accomplish this by producing a partition $\mathcal{P} = \{t_0, t_1, \ldots .t_n\}$ of $[a, b]$ with the property that

$$\lambda(t_j) - \lambda(t_{j-1}) < \frac{1}{2}\varepsilon \quad \text{for } j = 1, \ldots, n.$$

Then we'll know $|\lambda(s) - \lambda(t)| < \varepsilon$ whenever s and t are in either the same subinterval or adjacent subintervals defined by $\mathcal{P}$, and so our δ can be the length of the shortest subinterval.

For any choice of $\mathcal{P}$ and any j, the error in approximating $\lambda(t_j) - \lambda(t_{j-1})$ by $|P_{j-1}P_j|$ is never greater than the error in approximating the arc length of all of $\mathcal{C}$ by $\sum_{k=1}^n |P_{k-1}P_k|$, and so

$$\lambda(t_j) - \lambda(t_{j-1}) \leq |P_{j-1}P_j| + \left(\lambda(b) - \sum_{k=1}^n |P_{k-1}P_k|\right).$$

Consequently, all we need to do is make sure that every segment of the broken-line path is shorter than $\frac{1}{4}\varepsilon$ and that the total length of the broken-line path corresponding to $\mathcal{P}$ is within $\frac{1}{4}\varepsilon$ of the arc length of $\mathcal{C}$.

By Theorem 5.2 in Chapter 2, there is a $\delta_0 > 0$ such that every s and t in $[a, b]$ with $|s - t| < \delta_0$ satisfy

$$|f(s) - f(t)| < \frac{1}{8}\varepsilon \quad \text{and} \quad |g(s) - g(t)| < \frac{1}{8}\varepsilon.$$

So whenever the mesh of $\mathcal{P}$ is less than δ_0, every segment satisfies

$$|P_{k-1}P_k| = \sqrt{[f(t_k) - f(t_{k-1})]^2 + [g(t_k) - g(t_{k-1})]^2} < \frac{1}{4}\varepsilon.$$

The definition of arc length shows there is a partition with

$$\lambda(b) < \frac{1}{4}\varepsilon + \sum_{k=1}^n |P_{k-1}P_k|,$$

and if necessary we can insert additional points to produce such a partition with mesh smaller than δ_0. ∎

EXERCISES

1. We've proved earlier that an increasing function defined on an interval is continuous if its set of values is an interval. Why didn't we use that to prove Theorem 1.2?

2. Show that if f and g are continuous, monotonic numerical functions on $[a, b]$, then the curve defined by

$$P(t) = (f(t), g(t)) \quad \text{for } a \le t \le b$$

is rectifiable with arc length at most $|f(b) - f(a)| + |g(b) - g(a)|$.

3. Explain why considering partitions $\{t_0, t_1, \ldots, t_n\}$ of $[a, b]$ with all terms $|P_{k-1}P_k|$ arbitrarily small won't necessarily produce sums that approach the arc length of the curve defined by $P(t)$, even when the curve is rectifiable.

4. Suppose that the curve given by $P(t)$ for $a \le t \le b$ has arc length $(1 + \varepsilon)|P(a)P(b)|$, where ε is a small positive number. For $a \le t \le b$, how far can the point $P(t)$ be from the line segment $\overline{P(a)P(b)}$?

2 ARC LENGTH AND INTEGRATION

The arc-length function in the previous section has much in common with functions defined by integrals. In both cases we can approximate their values in terms of finite sums associated with partitions, and refining the partitions improves the accuracy of the approximations. This isn't just a coincidence; it is often possible to use an integral to calculate arc length, with the integrand derived from the way we represent the curve. The standard formula for finding the arc length of a curve given parametrically by

$$P(t) = (f(t), g(t)) \quad \text{for } a \le t \le b$$

is
$$L = \int_a^b \sqrt{f'(t)^2 + g'(t)^2} \, dt.$$

This has an appealing physical interpretation when we think of the parameter t as time. Then the derivatives $f'(t)$ and $g'(t)$ are the x- and y-components of the velocity of the moving point, and the integrand

$$v(t) = \sqrt{f'(t)^2 + g'(t)^2}$$

represents the magnitude of the velocity, better known as the speed. Thus the integral formula for arc length says that we integrate speed with respect to time to find the distance traveled. But there are some very real restrictions on the functions f and g that are not always met, even for very simple curves. Our task in studying this formula is twofold: we need conditions that assure the integrand is Riemann integrable, and we need to show that the value of the integral is indeed the arc length.

While we think of f' and g' as derivatives, what we really need from them is that their integrals give us f and g:

$$f(t) = f(a) + \int_a^t f'(\tau)\,d\tau \quad \text{and} \quad g(t) = g(a) + \int_a^t g'(\tau)\,d\tau.$$
(7.1)

Of course, that's the case when f and g are differentiable at all points in $[a,b]$ with f' and g' continuous, but (7.1) may be valid under more general conditions. The possible existence of points at which f or g is not differentiable is of no concern to us; we just need f' and g' to be Riemann integrable functions that satisfy (7.1). Once again calling $P_k = (f(t_k), g(t_k))$, we calculate

$$
\begin{aligned}
|P_{k-1}P_k| &= \left\{ [f(t_k) - f(t_{k-1})]^2 + [g(t_k) - g(t_{k-1})]^2 \right\}^{1/2} \\
&= \left\{ \left[\int_{t_{k-1}}^{t_k} f'(t)\,dt \right]^2 + \left[\int_{t_{k-1}}^{t_k} g'(t)\,dt \right]^2 \right\}^{1/2}.
\end{aligned}
$$

Thus the values of f' and g' determine how much the position of $P(t)$ can change over various subintervals of $[a,b]$. In particular, for f' and g' Riemann integrable over $[a,b]$ they must be bounded, and that lets us write

$$|P_{k-1}P_k| \le \sqrt{A^2 + B^2}\,(t_k - t_{k-1})$$

when $|f'(t)| \le A$ and $|g'(t)| \le B$ for all $t \in [a,b]$. Such a bound shows that $\mathcal{C}$ is rectifiable, with arc length at most $\sqrt{A^2 + B^2}\,(b-a)$.

Knowing that $\mathcal{C}$ is rectifiable simplifies our task; when f' and g' are Riemann integrable we can use Theorem 2.1 from Chapter 5 to show that v is Riemann integrable over $[a,b]$ and that the arc length L of $\mathcal{C}$ is the value of $\int_a^b v(t)\,dt$. That involves showing that for any $\varepsilon > 0$, there is a partition of $[a,b]$ such that every associated Riemann sum for v is within ε of L. As usual, we let $\mathcal{P} = \{t_k\}_{k=0}^n$ represent an arbitrary partition of $[a,b]$, and we let $\{t_k^*\}_{k=1}^n$ represent an arbitrary set of sampling points for $\mathcal{P}$. We then write

$$\sum_{k=1}^n v(t_k^*)\,\Delta t_k = \sum_{k=1}^n |P_{k-1}P_k| + \sum_{k=1}^n (v(t_k^*)\,\Delta t_k - |P_{k-1}P_k|). \quad (7.2)$$

Since $\mathcal{C}$ is rectifiable, we know we can make $\sum_{k=1}^n |P_{k-1}P_k|$ be within $\varepsilon/2$ of L by choosing $\mathcal{P}$ appropriately, so our job is to show that we can make the last sum in equation (7.2) be smaller than $\varepsilon/2$ as well. That requires a careful use of the Riemann integrability of f' and g'.

To help us make better use of these assumptions, let's call m_k and M_k the infimum and supremum of the values of f' over $[t_{k-1}, t_k]$, so that

$$L(\mathcal{P}) = \sum_{k=1}^{n} m_k \Delta t_k \quad \text{and} \quad U(\mathcal{P}) = \sum_{k=1}^{n} M_k \Delta t_k$$

are the lower and upper sums for f' relative to $\mathcal{P}$. Then the diameter of the set of Riemann sums for f' associated with the partition $\mathcal{P}$ is

$$U(\mathcal{P}) - L(\mathcal{P}) = \sum_{k=1}^{n} (M_k - m_k) \Delta t_k.$$

Let's also write m'_k, M'_k, $L'(\mathcal{P})$, and $U'(\mathcal{P})$ for the analogous quantities when f' is replaced by g'.

For each k we have

$$m_k \Delta t_k \leq \int_{t_{k-1}}^{t_k} f'(t)\, dt \leq M_k \Delta t_k,$$

so that

$$\xi_k = \frac{1}{\Delta t_k} \int_{t_{k-1}}^{t_k} f'(t)\, dt = \frac{f(t_k) - f(t_{k-1})}{\Delta t_k}$$

defines a number in $[m_k, M_k]$. Similarly,

$$\eta_k = \frac{g(t_k) - g(t_{k-1})}{\Delta t_k}$$

is a number in $[m'_k, M'_k]$, and we can write

$$|P_{k-1}P_k| = \left\{ [f(t_k) - f(t_{k-1})]^2 + [g(t_k) - g(t_{k-1})]^2 \right\}^{1/2}$$
$$= \sqrt{\xi_k^2 + \eta_k^2}\, \Delta t_k.$$

To bound the difference between $|P_{k-1}P_k|$ and $v(t_k^*) \Delta t_k$, we use an elementary inequality:

$$\left| \sqrt{a^2 + b^2} - \sqrt{c^2 + d^2} \right| \leq |a - c| + |b - d|. \tag{7.3}$$

To explain this inequality, call

$$O = (0,0), \ P = (a,b), \ Q = (c,d), \text{ and } R = (c,b).$$

The triangle inequality shows that

$$\left| |OP| - |OQ| \right| \leq |PQ| \leq |PR| + |RQ|,$$

and substituting for the lengths of these segments gives (7.3). So for each k we have

$$\left| v\left(t_k^*\right) \Delta t_k - |P_{k-1} P_k| \right| = \left| \sqrt{f'\left(t_k^*\right)^2 + g'\left(t_k^*\right)^2} - \sqrt{\xi_k^2 + \eta_k^2} \right| \Delta t_k$$
$$\leq \left(\left| f'\left(t_k^*\right) - \xi_k \right| + \left| g'\left(t_k^*\right) - \eta_k \right| \right) \Delta t_k$$
$$\leq \left[\left(M_k - m_k \right) + \left(M_k' - m_k' \right) \right] \Delta t_k.$$

Now we bound the last sum in (7.2). Since the absolute value of a sum is never more than the sum of the absolute values, we get

$$\left| \sum_{k=1}^{n} \left(v\left(t_k^*\right) \Delta t_k - |P_{k-1} P_k| \right) \right| \leq \sum_{k=1}^{n} \left[\left(M_k - m_k \right) + \left(M_k' - m_k' \right) \right] \Delta t_k$$
$$= U\left(\mathcal{P}\right) - L\left(\mathcal{P}\right) + U'\left(\mathcal{P}\right) - L'\left(\mathcal{P}\right).$$

That too can be made smaller than $\varepsilon/2$ by choosing $\mathcal{P}$ appropriately, and the same choice of $\mathcal{P}$ can also make $\sum_{k=1}^{n} |P_{k-1} P_k|$ be within $\varepsilon/2$ of L. The theorem below summarizes what we've shown.

THEOREM 2.1:   *Let C be the curve $P(t) = (f(t), g(t))$ for $a \leq t \leq b$, and suppose that f' and g' are Riemann integrable functions with (7.1) satisfied on $[a, b]$. Then C is rectifiable, and its arc length is*

$$L = \int_a^b \sqrt{f'(t)^2 + g'(t)^2} \, dt.$$

In particular, these conditions are satisfied when the derivatives of f and g are continuous on $[a, b]$.

As we use this theorem, we should keep in mind that it gives a sufficient condition for a curve to be rectifiable, not a necessary one. Many different pairs of functions can describe exactly the same curve, with the points traced in the same order, and the theorem may only apply to some of these representations. For example, the straight line segment

$$f(t) = g(t) = t, \ 0 \leq t \leq 1$$

is also given by

$$f(t) = g(t) = \sqrt{t}, \ 0 \le t \le 1;$$

the theorem applies to the first parametrization but not to the second. However, when two different parametrizations satisfy the hypotheses of the theorem, the resulting integrals must all have the same value, even though the integrands can be quite different.

EXERCISES

5. Parametrize the top of the unit circle by calling $(f(t), g(t))$ the point where the line segment from $(0,0)$ to $(t,1)$ crosses the unit circle. Find the function v we integrate to find arc length from $P(a)$ to $P(b)$; $\int v(t) \, dt$ can be reduced to a standard form. Does some choice of a and b make $\int_a^b v(t) \, dt$ correspond to the arc length of the entire top half of the circle?

6. Parametrize the unit circle by calling $(f(t), g(t))$ the point above the line $y = -1$ where the line through $(0, -1)$ and $(t, 0)$ intersects the unit circle. Find the function $v(t)$ we integrate to find arc length from $P(a)$ to $P(b)$ on the unit circle. (This parametrization is especially useful because f, g, and v are simple rational functions.) Does $\int_a^b v(t) \, dt$ ever correspond to the arc length of the top of the circle?

3 ARC LENGTH AS A PARAMETER

Now let's consider a slightly different class of curves. We again represent $\mathcal{C}$ in the form

$$P(t) = (f(t), g(t)) \quad \text{for } t \in I, \text{ with } f, g \in \mathcal{C}(I),$$

but this time we assume that I is an open interval, so that $\mathcal{C}$ has no first or last point. Instead of assuming that the entire curve is rectifiable, we suppose that for every closed subinterval $[a, b] \subset I$, the portion of $\mathcal{C}$ between $P(a)$ and $P(b)$ is rectifiable. Then there is still an increasing function λ defined on I with the property that $\lambda(b) - \lambda(a)$ gives the arc length of the portion of $\mathcal{C}$ between $P(a)$ and $P(b)$, even though $\lambda(t)$ need not be the arc length of the portion of $\mathcal{C}$ preceding $P(t)$. We simply pick a base point $t_0 \in I$ and specify $\lambda(t_0) = 0$. Then $\lambda(t)$ is positive for $t > t_0$ and negative for $t < t_0$; it's common to call $\lambda(t)$ the **directed arc length**. Instead of using t to identify different points on $\mathcal{C}$, in principle we can use $\lambda(t)$ just as well.

If f and g are functions in $\mathcal{C}^1(I)$, then all the assumptions we made in the previous section about the functions defining $\mathcal{C}$ are satisfied on each closed subinterval of I, so

$$\lambda(t) = \int_{t_0}^{t} \sqrt{f'(\tau)^2 + g'(\tau)^2}\, d\tau. \tag{7.4}$$

Since $\lambda(t)$ is the integral of a function that is continuous on I, Theorem 4.1 in Chapter 5 tells us that

$$\lambda'(t) = \sqrt{f'(t)^2 + g'(t)^2} \quad \text{for all } t \in I.$$

Thus λ is itself a function in $C^1(I)$. If we also assume that $f'(t)$ and $g'(t)$ do not vanish simultaneously, then $\lambda'(t) > 0$ for all $t \in I$. In that case, the inverse function theorem shows that the equation

$$s = \lambda(t)$$

defines t as a differentiable function of s. That is, there is a differentiable function T on an open interval J such that

$$T(\lambda(t)) = t \quad \text{for all } t \in I,$$

and its derivative satisfies

$$T'(s) = \frac{1}{\lambda'(t)} \quad \text{when } s = \lambda(t).$$

Of all the ways to describe the same curve, mathematicians favor expressing x and y as functions of the directed arc length. The reason is that the parameter s has an intrinsic geometric meaning determined by the points on $\mathcal{C}$ and the order in which they are encountered. Consequently, the derivatives $\frac{dx}{ds}$ and $\frac{dy}{ds}$ reflect the geometry of the curve in ways that $f'(t)$ and $g'(t)$ may not. In principle, other parametric representations can be converted to this form by finding $T(s)$ and substituting it for t:

$$x = f(T(s)), y = g(T(s)) \quad \text{for } s \in J.$$

In practice, we may not be able to find a usable formula for $T(s)$, but we can still find the derivatives $\frac{dx}{ds}$ and $\frac{dy}{ds}$ in terms of t by using the chain rule and our formula for $T'(s)$.

For example, let's consider a fairly arbitrary curve that follows the unit circle; that is, a curve described by

$$x = f(t), y = g(t) \quad \text{with } f(t)^2 + g(t)^2 = 1 \quad \text{for all } t \in I.$$

To be able to use our formulas, we'll assume that f and g are in $C^1 \, (I)$, and we'll further assume that

$$f \left(t \right) g' \left(t \right) - f' \left(t \right) g \left(t \right) > 0 \quad \text{for all } t \in I.$$

In addition to implying that f' and g' never vanish simultaneously, this also implies that the motion is counterclockwise because it makes $g \left(t \right) / f \left(t \right)$ increasing where $f \left(t \right) \neq 0$. We'll calculate $\frac{dx}{ds}$ and $\frac{dy}{ds}$ for this curve. By the way, all our assumptions are valid for

$$f \left(t \right) = \frac{1 - t^2}{1 + t^2} \quad \text{and} \quad g \left(t \right) = \frac{2t}{1 + t^2},$$

but our calculations are equally valid for many other parametrizations.

By differentiating the equation

$$f \left(t \right)^2 + g \left(t \right)^2 = 1$$

with respect to t, we quickly discover that

$$f \left(t \right) f' \left(t \right) = -g \left(t \right) g' \left(t \right).$$

Consequently, we may write

$$\begin{aligned} f' \left(t \right) &= \left[f \left(t \right)^2 + g \left(t \right)^2 \right] f' \left(t \right) \\ &= f \left(t \right) \left[-g \left(t \right) g' \left(t \right) \right] + g \left(t \right)^2 f' \left(t \right) \\ &= - \left[f \left(t \right) g' \left(t \right) - f' \left(t \right) g \left(t \right) \right] g \left(t \right) \end{aligned}$$

as well as

$$\begin{aligned} g' \left(t \right) &= \left[f \left(t \right)^2 + g \left(t \right)^2 \right] g' \left(t \right) \\ &= f \left(t \right)^2 g' \left(t \right) + g \left(t \right) \left[-f \left(t \right) f' \left(t \right) \right] \\ &= \left[f \left(t \right) g' \left(t \right) - f' \left(t \right) g \left(t \right) \right] f \left(t \right). \end{aligned}$$

The similarities in the formulas for $f' \left(t \right)$ and $g' \left(t \right)$ are a real help in calculating $\lambda' \left(t \right) = v \left(t \right)$. Since

$$\begin{aligned} f' \left(t \right)^2 + g' \left(t \right)^2 &= \left[f \left(t \right) g' \left(t \right) - f' \left(t \right) g \left(t \right) \right]^2 g \left(t \right)^2 \\ &\quad + \left[f \left(t \right) g' \left(t \right) - f' \left(t \right) g \left(t \right) \right]^2 f \left(t \right)^2 \\ &= \left[f \left(t \right) g' \left(t \right) - f' \left(t \right) g \left(t \right) \right]^2 \end{aligned}$$

and we've assumed that the quantity in brackets is positive, we have

$$\lambda'(t) = \sqrt{f'(t)^2 + g'(t)^2} = f(t)g'(t) - f'(t)g(t).$$

Consequently, we obtain the formulas

$$\frac{dx}{ds} = f'(T(s))T'(s) = \frac{f'(t)}{\lambda'(t)} = -g(t) = -y,$$

$$\frac{dy}{ds} = g'(T(s))T'(s) = \frac{g'(t)}{\lambda'(t)} = f(t) = x.$$

Formulas for $f'(t)$ and $g'(t)$ may not be nearly so simple; they depend strongly on the specific formulas for $f(t)$ and $g(t)$.

It's easy enough to imagine a curve that wraps around the circle infinitely often, and the use of arc length to parametrize this curve should be quite familiar. Mathematicians prefer to orient the curve in the counterclockwise direction, with $s = 0$ corresponding to the point $(1, 0)$. Then its parametrization in terms of directed arc length is

$$x = \cos s, \; y = \sin s \quad \text{for} \; -\infty < s < \infty,$$

and this is taken as the definition of the circular functions *cosine* and *sine*. For $0 < s < \frac{\pi}{2}$, these agree with the usual trigonometric ratios corresponding to an acute angle with radian measure s, so the terms *circular functions* and *trigonometric functions* are used almost interchangeably. The notation for the trigonometric functions was standardized before calculus notation developed, which is why we usually write $\sin s$ and $\cos s$ instead of $\sin(s)$ and $\cos(s)$. The formulas

$$\frac{d}{ds}(\cos s) = -\sin s \quad \text{and} \quad \frac{d}{ds}(\sin s) = \cos s$$

are just restatements of the formulas for $\frac{dx}{ds}$ and $\frac{dy}{ds}$ that we derived above.

For more general curves, the derivatives $\frac{dx}{ds}$ and $\frac{dy}{ds}$ are used to determine where the curve is smooth; other derivatives may not provide the right information. For example, the graph of $y = f(x)$ is certainly smooth where $f'(x)$ is continuous, but it can also be smooth where f isn't even differentiable; a smooth curve can have a vertical tangent line. Or when a curve is expressed in the form $x = f(t)$, $y = g(t)$, the curve may not be smooth at all points where $f'(t)$ and $g'(t)$ are continuous; it can have a corner at a point where both derivatives vanish. The exercises below illustrate some of these ideas.

EXERCISES

7. The parametrizations of the unit circle found in the previous set of exercises describe clockwise motion around the unit circle. By simply computing all the indicated quantities, show that

$$f'(t) = g(t)\lambda'(t) \quad \text{and} \quad g'(t) = -f(t)\lambda'(t).$$

8. The graph of $y = x^{1/3}$ can be parametrized by $x = t^3$, $y = t$ for t in the interval $(-\infty, \infty)$. Without actually expressing x and y as functions of directed arc length, find $\frac{dx}{ds}$ and $\frac{dy}{ds}$ as functions of t and show that they are continuous.

9. The graph of $x = t^3$, $y = t^2$ has a cusp at the origin. Show that $\frac{dy}{ds}$ is discontinuous there by considering the cases $t > 0$ and $t < 0$ separately.

10. The graph of $x = t^3$, $y = t|t|$ is a smooth curve. Show that $\frac{dy}{ds}$ is continuous at the origin by considering the cases $t > 0$ and $t < 0$ separately.

4   THE ARCTANGENT AND ARCSINE FUNCTIONS

Although the trigonometric functions are more familiar to us, in many ways the inverse trigonometric functions are simpler. Arc length is basically hard to measure, and locating a point corresponding to a specified arc length is harder still. We know exact values for the circular functions at multiples of $\frac{\pi}{4}$, for example, because we recognize $\frac{\pi}{4}$ as exactly one-eighth the circumference of the unit circle and we can use geometric constructions to locate the appropriate point on the unit circle. But to locate the point $(\cos 1, \sin 1)$ with any degree of precision, some heavy-duty calculations are involved. Since we now do those things by just pushing a few buttons on a calculator, it's easy to lose sight of how much is really going on. It's fundamentally simpler to locate a point with given coordinates and then determine arc length up to that point than it is to determine the coordinates of a point corresponding to a given arc length. That makes it easier to develop methods for calculating inverse trigonometric functions.

The simplest of the inverse trigonometric functions is the arctangent. For $P(t)$ the point where the line segment from the origin to $(1, t)$ intersects the unit circle, $\arctan(t)$ is the directed arc length from $P(0)$ to $P(t)$. We'll use this geometric description of $P(t)$ to find a formula for a curve. Then we'll use equation (7.4) to produce an integral for the directed arc length.

The formula for $P(t)$ is simple enough:

$$P(t) = \left(\frac{1}{\sqrt{1+t^2}}, \frac{t}{\sqrt{1+t^2}} \right).$$

So we call

$$f(t) = \frac{1}{\sqrt{1+t^2}} \quad \text{and} \quad g(t) = \frac{t}{\sqrt{1+t^2}} \quad \text{for } -\infty < t < \infty,$$

and note that f and g are in $\mathcal{C}^1(\mathbf{R})$.

To calculate arc length, we'll need some derivatives. We find

$$f'(t) = -\frac{t}{[1+t^2]^{3/2}}$$

and

$$g'(t) = \frac{1}{[1+t^2]^{1/2}} - \frac{t^2}{[1+t^2]^{3/2}} = \frac{1}{[1+t^2]^{3/2}}.$$

Thus

$$f'(t)^2 + g'(t)^2 = \frac{t^2}{[1+t^2]^3} + \frac{1}{[1+t^2]^3} = \frac{1}{[1+t^2]^2}.$$

Consequently, our integral formula for the arctangent is

$$\arctan(t) = \int_0^t \sqrt{f'(\tau)^2 + g'(\tau)^2}\, d\tau$$

$$= \int_0^t \frac{d\tau}{1+\tau^2} \quad \text{for all } t \in \mathbf{R}.$$

That also proves that

$$\frac{d}{dt}[\arctan(t)] = \frac{1}{1+t^2},$$

one of the standard formulas for derivatives.

The arcsine function is slightly more complicated than the arctangent because it is only defined on $[-1, 1]$. Geometrically, $\arcsin(t)$ is also defined to be the directed arc length from $P(0)$ to $P(t)$ along the unit circle, where $P(t)$ is the intersection of the horizontal line $y = t$ with the right half of the unit circle. This time we find that

$$P(t) = \left(\sqrt{1-t^2}, t \right).$$

Calling

$$f(t) = \sqrt{1 - t^2} \quad \text{and} \quad g(t) = t,$$

we note that

$$f'(t) = -\frac{t}{\sqrt{1 - t^2}} \quad \text{for } t \in (-1, 1).$$

While f' is continuous on $(-1, 1)$, it's not integrable over $[-1, 1]$ so we're only able to write $\arcsin(t)$ as an integral for $|t| < 1$. For those t we calculate

$$f'(t)^2 + g'(t)^2 = \frac{t^2}{1 - t^2} + 1 = \frac{1}{1 - t^2}.$$

Hence

$$\arcsin(t) = \int_0^t \frac{d\tau}{\sqrt{1 - \tau^2}} \quad \text{for } -1 < t < 1.$$

Since the circle is a rectifiable curve, we know that $\arcsin(t)$ is actually continuous on $[-1, 1]$, and we know that

$$\arcsin(1) = \frac{\pi}{2},$$

because the arc involved is a quarter-circle. We can conclude that

$$\lim_{t \to 1-} \left(\int_0^t \frac{d\tau}{\sqrt{1 - \tau^2}} \right) = \lim_{t \to 1-} \arcsin(t) = \frac{\pi}{2}.$$

This is often indicated in the condensed form

$$\int_0^1 \frac{d\tau}{\sqrt{1 - \tau^2}} = \frac{\pi}{2},$$

even though the function to be integrated is not Riemann integrable over $[0, 1]$. Such expressions are called **improper integrals**. Even though the notation is the same as for ordinary integrals, there are important differences. For example, Riemann sums do not provide reliably accurate approximations to improper integrals, no matter how small the mesh of the partition used.

It's also possible to calculate $\arctan(t)$ or $\arcsin(t)$ geometrically, in much the same way as we might calculate the arc length of the entire circle. We can form successive approximations as lengths of broken-line paths in a way that lets us calculate their lengths recursively. But we need not pursue that here.

EXERCISES

11. How can our integral formula for the arctangent be used to calculate π?

12. Show geometrically that

$$\arcsin (t) = \arctan \left(\frac{t}{\sqrt{1 - t^2}} \right) \quad \text{for } -1 < t < 1.$$

13. For $t > 1$, let $P(t)$ be the first-quadrant point on the unit circle from which the tangent line passes through $(t, 0)$. Find the function $v(t)$ we integrate to find directed arc length along this curve. What's the problem in using the corresponding integral to define the arcsecant?

5   THE FUNDAMENTAL TRIGONOMETRIC LIMIT

Our use of equation (7.4) and the inverse function theorem to derive the formulas

$$\frac{d}{ds} (\sin s) = \cos s \quad \text{and} \quad \frac{d}{ds} (\cos s) = -\sin s$$

is certainly not the way it's done in introductory calculus courses. We had the luxury of not needing the formulas quickly. They're ordinarily developed by using trigonometric identities to express the needed limits in terms of the fundamental trigonometric limit

$$\lim_{s \to 0} \frac{\sin s}{s} = 1.$$

This limit is closely related to our understanding of arc length on the unit circle. Each side of an inscribed regular n-sided polygon subtends an angle $2\pi/n$, so each side has length $2 \sin (\pi/n)$. Consequently,

$$2\pi = \lim_{n \to \infty} 2n \sin (\pi/n),$$

and we can rewrite that formula as

$$\lim_{n \to \infty} \frac{\sin (\pi/n)}{\pi/n} = 1.$$

That's strong evidence for the fundamental limit, but it doesn't fully establish it because every interval $(-\delta, \delta)$ will contain lots of numbers that aren't of the form π/n. Just a little more work is needed. Symmetry shows that we only need to consider $s > 0$, and we note that

$$\sin\left[\pi/\left(n+1\right)\right] \le \sin s \le \sin\left(\pi/n\right) \quad \text{when } \pi/\left(n+1\right) \le s \le \pi/n.$$

So for all such s we have

$$\frac{\sin\left[\pi/\left(n+1\right)\right]}{\pi/n} \le \frac{\sin s}{s} \le \frac{\sin\left(\pi/n\right)}{\pi/\left(n+1\right)},$$

and we can rewrite this as

$$\frac{n}{n+1} \cdot \frac{\sin\left[\pi/\left(n+1\right)\right]}{\pi/\left(n+1\right)} \le \frac{\sin s}{s} \le \frac{n+1}{n} \cdot \frac{\sin\left(\pi/n\right)}{\pi/n}.$$

Then as s decreases to 0, $n \to \infty$ and that shows $\left(\sin s\right)/s \to 1$.

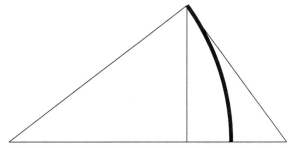

Figure 7.2　Comparison of s, $\sin s$, and $\tan s$.

Most calculus books establish the fundamental limit as a consequence of the inequality

$$\sin s < s < \tan s \quad \text{for all } s \in \left(0, \frac{\pi}{2}\right). \tag{7.5}$$

That certainly agrees with visual estimates of the distances in Figure 7.2, where an arc of length s on the unit circle is between a vertical line of length $\sin s$ and a slanted line of length $\tan s$. To measure s we would calculate either $\arcsin\left(\sin s\right)$ or $\arctan\left(\tan s\right)$, and the integral formulas we developed in the previous section show that for $0 < s < \frac{\pi}{2}$ we have

$$\sin s < \arcsin\left(\sin s\right) \quad \text{and} \quad \arctan\left(\tan s\right) < \tan s.$$

Inequality (7.5) also follows from the ideas in our first discussion of arc length on circles, if we double all the terms by adding a bottom half that is the mirror image of the top half. That makes $2s$ the length of an arc subtending a chord of length $2 \sin s$, and the ends of the arc are joined by a broken-line path of length $2 \tan s$ that is part of a circumscribed polygon.

Arc length is a subtle enough concept that most calculus books avoid worrying about it too much. In the 1960s it became fashionable to establish (7.5) by comparing areas instead. The ratio of the area of a circular sector determined by an arc of length s to the area of the whole circle should be the same as the ratio of s to the circumference of the circle. So since the whole unit circle has area π, the sector must have area $\frac{1}{2}s$. But this sector obviously contains a triangle of area $\frac{1}{2}\sin s$ and is contained in another triangle of area $\frac{1}{2}\tan s$, and that proves (7.5). However, this discussion ignores the question of why the arc length and area of a circle involve the same mysterious constant we call π. The concept of arc length really is an important part of every geometric approach to the fundamental trigonometric limit.

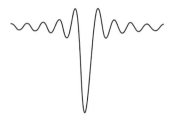

VIII

SEQUENCES AND SERIES OF FUNCTIONS

I n this chapter we'll study some additional ways to produce new functions and learn how we can work with the functions that result. These techniques are fairly general and have proved to be a valuable source of functions that meet needs of engineers and scientists as well as mathematicians.

I FUNCTIONS DEFINED BY LIMITS

Calculus operations usually involve limits in some sense, and the simplest sort of limit is the limit of a sequence. In Chapter 3 we proved things about limits of sequences of numbers. Now we'll consider limits of sequences of numerical functions. Generally, we'll assume that we have an interval I and an infinite sequence $\{f_n\}_{n=1}^{\infty}$ of numerical functions whose domains all include I. Then for each $x \in I$ the sequence $\{f_n(x)\}_{n=1}^{\infty}$ is a sequence of numbers. When each of those sequences is convergent, the formula

$$f(x) = \lim_{n \to \infty} f_n(x)$$

defines a numerical function f on I; we say that f is the **pointwise limit** of the sequence $\{f_n\}_{n=1}^{\infty}$. For example, the natural exponential function is the pointwise limit of the sequence $\{E_n\}_{n=1}^{\infty}$ given by

$$E_n(x) = \left(1 + \frac{1}{n}x\right)^n \quad \text{for} \ -\infty < x < \infty,$$

and the natural logarithm is the pointwise limit of the sequence $\{L_n\}_{n=1}^{\infty}$ defined by

$$L_n(x) = n\left(\sqrt[n]{x} - 1\right) \quad \text{for } x > 0.$$

In Chapter 6 we used other definitions for the natural logarithm and exponential function, then developed these sequences to provide useful approximations. Here we're concerned with using sequences both to define functions and to develop their properties, and that raises a different set of questions. How do we tell whether a given sequence of functions has a pointwise limit? When it does, how can we establish the continuity, differentiability, or integrability of the function it defines? How can we calculate the derivative or the integral? Ideally, we would like a simple way to answer such questions by inspecting the terms in the sequence used for the definition.

Determining whether a given sequence of functions has a pointwise limit is itself a challenge. In the case of the natural exponential function and the natural logarithm, we used properties of those functions to explain why the sequences $\{E_n(x)\}_{n=1}^{\infty}$ and $\{L_n(x)\}_{n=1}^{\infty}$ were convergent. But if the pointwise limit of a sequence of functions $\{f_n\}_{n=1}^{\infty}$ is to serve as the definition of f, then we can't count on using properties of f or even the value of $f(x)$ to help show why $\{f_n(x)\}_{n=1}^{\infty}$ converges. That's a serious problem since the limit of a sequence is defined in terms of a condition it must satisfy instead of as the result of performing some operation on the sequence. To get past this difficulty, we go back to the ideas of Chapters 1 and 3.

Recall that we first investigated the convergence of a bounded sequence $\{x_n\}_{n=1}^{\infty}$ of real numbers by considering the tails, defining

$$a_m = \inf\{x_n : n \geq m\} \quad \text{and} \quad b_m = \sup\{x_n : n \geq m\}.$$

That makes $\{[a_m, b_m]\}_{m=1}^{\infty}$ a nested sequence of closed intervals, and the sequence $\{x_n\}_{n=1}^{\infty}$ converges if and only if $b_m - a_m \to 0$. When the sequence does converge, the limit is the one number in $\bigcap_{m=1}^{\infty}[a_m, b_m]$.

But we don't need to find the sequence of intervals to tell whether $b_m - a_m \to 0$; a very simple test was developed in the early nineteenth century by Augustin Cauchy, a French mathematician. He observed that a sequence $\{x_n\}_{n=1}^{\infty}$ of real numbers has a limit if and only if for each given $\varepsilon > 0$ there is an integer N such that $|x_n - x_m| < \varepsilon$ for all $n, m \geq N$. Nowadays we say that sequences with this last property satisfy the **Cauchy condition**. Clearly every convergent sequence satisfies the Cauchy condition; when both n and m are large enough to put x_n and x_m within $\frac{1}{2}\varepsilon$ of the limit, the distance between x_n and x_m must be less than ε. On the other hand, we know in general that

$$b_N - a_N = \operatorname{diam}\{x_n : n \geq N\} = \sup\{|x_n - x_m| : n, m \geq N\}.$$

So if $|x_n - x_m| < \varepsilon$ for all $n, m \geq N$, then $0 \leq b_N - a_N \leq \varepsilon$. Consequently, when Cauchy's condition is satisfied we can prove $b_m - a_m \to 0$, and therefore we can use $\{[a_m, b_m]\}_{m=1}^{\infty}$ to define the limit of $\{x_n\}_{n=1}^{\infty}$. The fact that every sequence satisfying Cauchy's condition is convergent is one more version of the completeness property of **R**.

Of course, the pointwise limit of a sequence of functions is a somewhat more complicated matter because we have to deal with a different sequence $\{f_n(x)\}_{n=1}^{\infty}$ for each x. But there are no new ideas involved, just more details to take care of. To see if the sequence has a pointwise limit, we check the **pointwise Cauchy condition**; that is, we check whether it satisfies the Cauchy condition at each point. Explicitly, the sequence of numerical functions $\{f_n\}_{n=1}^{\infty}$ has a pointwise limit on I if for each $x \in I$ and each $\varepsilon > 0$ there is an integer N such that $|f_n(x) - f_m(x)| < \varepsilon$ for all $n, m \geq N$. Note that in this condition, the integer N is allowed to depend on x as well as ε; that can be important. But for now, the most important lesson is that when $\{f_n\}_{n=1}^{\infty}$ satisfies the pointwise Cauchy condition on I, then calling

$$f(x) = \lim_{n \to \infty} f_n(x), \quad x \in I$$

defines a function f on I that is the pointwise limit of the sequence.

It's also common to use an **infinite series** to define a function. While the terms *sequence* and *series* are almost interchangeable in everyday speech, they have quite different meanings in mathematics. To a mathematician, an infinite series always refers to a **sum** of infinitely many terms. The expression $\sum_{n=1}^{\infty} a_n$ is interpreted as shorthand for the sequence $\{s_n\}_{n=1}^{\infty}$ of **partial sums** defined by the formula $s_n = \sum_{k=1}^{n} a_k$. It's sometimes helpful to remember that the partial sums can be defined recursively by

$$s_1 = a_1 \quad \text{and} \quad s_{n+1} = s_n + a_{n+1} \quad \text{for all } n \in \mathbf{N}.$$

It's also common to use $\sum_{n=1}^{\infty} a_n$ to indicate $\lim_{n \to \infty} s_n$ when this sequence converges. When we say an infinite series is convergent, that's always understood to mean that the corresponding sequence of partial sums is convergent, so there's little need to develop a separate theory of functions defined by infinite series. Note that when

$$s_n(x) = \sum_{k=1}^{n} a_k(x) \quad \text{for all } n \in \mathbf{N} \text{ and all } x \in I,$$

questions about the continuity, differentiability, or integrability of the functions in the sequence $\{s_n\}_{n=1}^{\infty}$ can be answered by examining the functions in $\sum_{n=1}^{\infty} a_n$ one at a time.

When f is the pointwise limit of $\{f_n\}_{n=1}^{\infty}$, it's tempting to conjecture that f' is the pointwise limit of $\{f'_n\}_{n=1}^{\infty}$. This is true in many important cases; let's look at a couple of them. For the sequence giving the natural exponential function, we easily calculate

$$E'_n(x) = n\left(1 + \frac{1}{n}x\right)^{n-1} \frac{d}{dx}\left(1 + \frac{1}{n}x\right)$$

$$= \left(1 + \frac{1}{n}x\right)^{n-1}$$

$$= E_n(x) \bigg/ \left(1 + \frac{1}{n}x\right) \quad \text{when } n \neq -x.$$

Since $1 + \frac{1}{n}x \to 1$, we see that $E'_n(x) \to \exp(x)$ for each $x \in \mathbf{R}$, which is the derivative of $\exp(x)$. Similarly, we can also calculate

$$L'_n(x) = n \cdot \frac{1}{n}x^{1/n-1} = x^{1/n-1}.$$

For $x > 0$ we have $x^{1/n} \to 1$, so $L'_n(x) \to x^{-1}$, the derivative of $\log x$.

For functions defined by limits, making conjectures about their properties is an easy matter. But verifying those conjectures is another matter entirely. Some simple examples point up the need for supplementary hypotheses. If we define f_n on $[0,1]$ by $f_n(x) = x^n$, then it's easy to recognize that

$$\lim_{n\to\infty} f_n(x) = \begin{cases} 0, & 0 \leq x < 1 \\ 1, & x = 1. \end{cases}$$

Each f_n is continuous and strictly increasing on $[0,1]$, yet the pointwise limit is constant on $[0,1)$ and discontinuous at 1. Or if we define g_n on $\mathbf{R}$ by

$$g_n(x) = \frac{1}{n}\sin n^2 x,$$

then clearly $g_n(x) \to 0$ for each $x \in \mathbf{R}$. In this case the pointwise limit certainly defines a differentiable function, but since $g'_n(x) = n \cos n^2 x$, we see little relationship between the sequence of derivatives and the derivative of the pointwise limit.

Now let's look at an example that is not at all simple, a type of function first studied in the nineteenth century by the German mathematician Karl Weierstrass. He found an infinite series of differentiable functions converging to a continuous function that isn't differentiable at any point. That was a real shock to the mathematics community; virtually no one had thought such a thing was possible. It showed the need for developing the ideas of calculus carefully, because intuition could be wrong. We'll conclude this section by analyzing the function defined by

$$f(x) = \sum_{n=0}^{\infty} 4^{-n} \sin 4^{2n} x, \quad -\infty < x < \infty.$$

We'll show that there are no points at which it is differentiable and no intervals on which it is monotonic, even though it is continuous at all points.

First let's see why the given series converges for all $x \in \mathbf{R}$ and defines a continuous function. Since the series starts with $n = 0$ instead of $n = 1$, calling $s_n(x)$ the sum of the first n terms in the series makes

$$s_n(x) = \sum_{k=0}^{n-1} 4^{-k} \sin 4^{2k} x.$$

Then for $m < n$ we have

$$|s_n(x) - s_m(x)| = \left| \sum_{k=m}^{n-1} 4^{-k} \sin 4^{2k} x \right|$$

$$\leq \sum_{k=m}^{n-1} 4^{-k} = \frac{1}{3} \left(4^{1-m} - 4^{1-n} \right).$$

So we have $|s_n(x) - s_m(x)| < \frac{1}{3} \cdot 4^{1-N}$ for all $m, n \geq N$, which shows that our series satisfies the pointwise Cauchy condition on all of $\mathbf{R}$. Moreover, the limit function f satisfies

$$|f(x) - s_m(x)| \leq \frac{1}{3} \cdot 4^{1-m} \quad \text{for all } x \in \mathbf{R} \text{ and all } m \in \mathbf{N}.$$

Now let's prove that f is continuous at each point $a \in \mathbf{R}$. Our task is to show that when $\varepsilon > 0$ is given, there is a $\delta > 0$ such that every

$x \in (a - \delta, a + \delta)$ satisfies $|f(x) - f(a)| < \varepsilon$. The key is that f is almost the same as s_m for any large enough value of m, and each s_m is continuous since it is a finite linear combination of continuous functions. For any m, we have

$$
\begin{aligned}
|f(x) - f(a)| &= |f(x) - s_m(x) + s_m(x) - s_m(a) + s_m(a) - f(a)| \\
&\leq |f(x) - s_m(x)| + |s_m(x) - s_m(a)| + |s_m(a) - f(a)| \\
&\leq \frac{1}{3} \cdot 4^{1-m} + |s_m(x) - s_m(a)| + \frac{1}{3} \cdot 4^{1-m}.
\end{aligned}
$$

So we pick m large enough to make $4^{1-m} < \varepsilon$, and since s_m is continuous at a there must be a $\delta > 0$ such that every $x \in (a - \delta, a + \delta)$ satisfies $|s_m(x) - s_m(a)| < \frac{1}{3}\varepsilon$. That's the δ we need.

Now we prove that f is never differentiable by using some properties of the sine function. Since $\sin\left(x \pm \frac{\pi}{2}\right) = \pm \cos x$ and

$$
(\sin x + \cos x)^2 + (\sin x - \cos x)^2 = 2\sin^2 x + 2\cos^2 x = 2,
$$

either $\sin\left(x + \frac{\pi}{2}\right)$ or $\sin\left(x - \frac{\pi}{2}\right)$ must differ from $\sin x$ by at least 1. So with a fixed and m any nonnegative integer, if $4^{2m}h_m = \pm\frac{\pi}{2}$ and we choose the sign of h_m correctly, we get

$$
\left|\sin 4^{2m}(a + h_m) - \sin 4^{2m}a\right| \geq 1.
$$

We also have

$$
\sin 4^{2n}(a + h_m) = \sin 4^{2n}a \quad \text{for all } n > m.
$$

This last equation lets us write $f(a + h_m) - f(a)$ as the difference of two finite sums; we regroup them as

$$
\begin{aligned}
f(a + h_m) - f(a) &= 4^{-m}\left[\sin 4^{2m}(a + h_m) - \sin 4^{2m}a\right] \\
&\quad + s_m(a + h_m) - s_m(a).
\end{aligned}
$$

Hence

$$
|f(a + h_m) - f(a)| \geq 4^{-m} - |s_m(a + h_m) - s_m(a)|,
$$

and we note that $4^{-m} = 4^m \cdot \frac{2}{\pi}|h_m|$.

Now we bound $|s_m(a + h_m) - s_m(a)|$ by using the mean value theorem. There must be a number c_m between a and $a + h_m$ with

$$
s_m(a + h_m) - s_m(a) = h_m s_m'(c_m)
$$

and we calculate

$$\left| s_m' \left(c_m \right) \right| = \left| \sum_{k=0}^{m-1} 4^k \cos 4^{2k} c_m \right| \leq \sum_{k=0}^{m-1} 4^k < \frac{1}{3} \cdot 4^m.$$

So we obtain

$$\left| f \left(a + h_m \right) - f \left(a \right) \right| > 4^m \left(\frac{2}{\pi} - \frac{1}{3} \right) \left| h_m \right|.$$

That shows why $f' \left(a \right)$ can't exist. Even though $h_m \to 0$ as $m \to \infty$, we clearly have

$$\lim_{m \to \infty} \left| \frac{f \left(a + h_m \right) - f \left(a \right)}{h_m} \right| = \infty.$$

Similar calculations explain why f is nowhere monotonic. Given a non-trivial interval I, when m is large enough the interval I will have a closed subinterval I_m of length $4^{-2m}\pi$ that is centered at a point a_m with $\sin 4^{2m} a_m = 1$. Then for each $n \geq m$ the term $\sin 4^{2n} x$ will vanish at both ends of I_m, and for $n > m$ it will vanish at a_m as well. The changes in s_m over I_m are small enough to guarantee that $f \left(a_m \right)$ is strictly larger than the value of f at either endpoint of I_m.

In Chapter 2 we mentioned that we have no procedure for locating the extrema of an arbitrary continuous function; this example gives a hint of the difficulties that can be involved. When we can find extreme values, we usually do it by using derivatives to split the interval into subintervals on which the function is monotonic, and then compare the values of the function at the endpoints of the subintervals. Here that's entirely out of the question.

EXERCISES

1. For f_n defined on $\mathbf{R}$ by the formula

$$f_n \left(x \right) = \frac{n^2 x^2}{n^2 x^2 + 1},$$

let f be the pointwise limit of $\{f_n\}_{n=1}^{\infty}$ on $\mathbf{R}$. Graph f. Where is f discontinuous?

2. Define $f_n \left(x \right) = n\phi \left(nx \right)$, where

$$\phi \left(x \right) = \begin{cases} 1 - |x - 1|, & |x - 1| \leq 1 \\ 0, & |x - 1| > 1. \end{cases}$$

Show that $f_n(x) \to 0$ for all $x \in \mathbf{R}$, but that

$$\int_0^2 f_n(x)\, dx = 1 \quad \text{for all } n.$$

2 CONTINUITY AND UNIFORM CONVERGENCE

Now let's develop a general condition under which we can prove that the pointwise limit of continuous functions is itself a continuous function. We'll assume that each function f_n in the sequence $\{f_n\}_{n=1}^{\infty}$ is defined on a fixed interval I, and that $f_n(x) \to f(x)$ for all $x \in I$. How do we prove that f is continuous at a point $a \in I$? Given $\varepsilon > 0$, we need to be able to produce $\delta > 0$ such that every $x \in I \cap (a - \delta, a + \delta)$ satisfies $|f(x) - f(a)| < \varepsilon$. Since we usually know more about the functions f_n in the sequence than we know about the limit f, we need to use the inequality

$$|f(x) - f(a)| \le |f(x) - f_n(x)| + |f_n(x) - f_n(a)| + |f_n(a) - f(a)|.$$

If f is the pointwise limit of the sequence, then for each $x \in I$ we can find an n such that both

$$|f(x) - f_n(x)| < \frac{1}{3}\varepsilon \quad \text{and} \quad |f(a) - f_n(a)| < \frac{1}{3}\varepsilon,$$

and if f_n is continuous at a then we can find a $\delta > 0$ such that every x in the interval $I \cap (a - \delta, a + \delta)$ satisfies

$$|f_n(x) - f_n(a)| < \frac{1}{3}\varepsilon$$

as well. So what's the problem? Doesn't this prove that f is continuous at a?

In fact, we've got a very big problem. In general we need the value of x to tell us what n to use, and when we use the continuity of f_n at a to determine our δ, that means we can't choose x until we've chosen n. It's a classic example of circular reasoning, but even if we don't recognize the logical fallacy we should recognize that something must be wrong somewhere. After all, we've seen an example in which the pointwise limit of a sequence of continuous functions is discontinuous.

In the previous section we used a standard method to establish the continuity of our nowhere differentiable function. We were able to choose m with $|f(x) - s_m(x)| < \varepsilon/3$ for all x, rather than having to choose an m that depended on any particular x under consideration. Choosing m

independently of x lets us use the continuity of the corresponding function in the sequence to prove the continuity of the limit. The definition and theorem below give a systematic way to exploit this simple idea.

DEFINITION 2.1:   Suppose that $\{f_n\}_{n=1}^{\infty}$ is a sequence of numerical functions all defined on I, and $f_n \to f$ pointwise on I. We say that $\{f_n\}_{n=1}^{\infty}$ **converges uniformly** to f on I if for each $\varepsilon > 0$ there is an integer N such that every function f_n with $n \geq N$ satisfies

$$|f_n(x) - f(x)| < \varepsilon \quad \text{for all } x \in I.$$

We often indicate this condition by writing $f_n \to f$ *uniformly* or by saying that f is the **uniform limit** of the sequence of functions. Note that pointwise convergence is a consequence of the last condition in the definition of uniform convergence, so we need not check for it separately. Of course, when pointwise convergence fails, uniform convergence must fail as well. Once we understand how uniform convergence differs from pointwise convergence, it's a simple matter to prove the theorem below. The steps in proving it are all described above, so we'll leave a formal proof to the reader.

THEOREM 2.1:   *Let $\{f_n\}_{n=1}^{\infty}$ be a sequence of numerical functions all defined on a fixed interval I, and suppose that $f : I \to \mathbf{R}$ is the uniform limit of $\{f_n\}_{n=1}^{\infty}$ on I. Then f is continuous at each point where all the functions in the sequence are continuous.*

One way to prove that a sequence $\{f_n\}_{n=1}^{\infty}$ of functions on an interval I is uniformly convergent is to show that it satisfies a **uniform Cauchy condition** on I. That is, for each $\varepsilon > 0$, there is an integer N such that $|f_n(x) - f_m(x)| < \varepsilon$ for all $n, m \geq N$ and for all $x \in I$. That's the route we took to establish uniform convergence for $\sum_{n=0}^{\infty} 10^{-n} \sin 10^{2n}x$, for example. Let's see why that works. Of course, if a sequence satisfies a uniform Cauchy condition on an interval, then it also satisfies a pointwise Cauchy condition on the same interval, and so the sequence must converge pointwise to a limit function f. But to prove uniform convergence, whenever $\varepsilon > 0$ is given we need to show there is an integer N such that $|f_n(x) - f(x)| < \varepsilon$ for all $n \geq N$ and all $x \in I$. We note that for every $m, n \in \mathbf{N}$ and every $x \in I$ we have

$$|f_n(x) - f(x)| \leq |f_n(x) - f_m(x)| + |f_m(x) - f(x)|.$$

Thus if $\{f_n\}_{n=1}^{\infty}$ satisfies a uniform Cauchy condition on I, then there is an N such that

$$|f_n(x) - f(x)| < \frac{\varepsilon}{2} + |f_m(x) - f(x)|$$

for all $n, m \geq N$ and all $x \in I$. Since $f_n \to f$ pointwise on I, for each $x \in I$ there must be an integer $m \geq N$ with $|f_m(x) - f(x)| < \varepsilon/2$. Therefore,

$$|f_n(x) - f(x)| < \frac{\varepsilon}{2} + \frac{\varepsilon}{2} = \varepsilon$$

for all $x \in I$ and all $n \geq N$.

To prove that an infinite series of functions is uniformly convergent, we use the same idea but the calculations look somewhat different. Of course, when we say $\sum_{n=1}^{\infty} f_n(x)$ converges uniformly, that means that the sequence $\{S_n(x)\}_{n=1}^{\infty}$ defined by

$$S_n(x) = \sum_{k=1}^{n} f_k(x)$$

converges uniformly, and we can still show this by proving the sequence satisfies a uniform Cauchy condition. But in the formula for $S_n(x) - S_m(x)$ we can always cancel the common terms in the sums, and that gives a different look to our uniform Cauchy condition. Given $\varepsilon > 0$, we need to be able to find an integer N such that

$$\left| \sum_{k=m+1}^{n} f_k(x) \right| < \varepsilon \quad \text{for all } x \in I \text{ and all } m, n \text{ with } N \leq m < n.$$

A moment's thought shows that we don't really change the usual Cauchy condition by considering only $N \leq m < n$ instead of all $m, n \geq N$, and in the case of series this change simplifies the notation significantly.

In many cases we can establish the uniform convergence of an infinite series of functions by using a simple condition known as the **Weierstrass M-test**. The theorem below gives a formal statement of just what it involves.

THEOREM 2.2: *Let $\{f_n\}_{n=1}^{\infty}$ be a sequence of real-valued functions whose domains all include a given set E, and let $\{M_n\}_{n=1}^{\infty}$ be a sequence of constants such that $|f_n(x)| \leq M_n$ for all $n \in \mathbf{N}$ and all $x \in E$. If $\sum_{n=1}^{\infty} M_n$ is convergent, then $\sum_{n=1}^{\infty} f_n(x)$ converges uniformly on E.*

☐ *Proof:* There's nothing to prove if $E = \emptyset$, and when $E \neq \emptyset$ we have a simple way to choose an N such that

$$\left| \sum_{k=m+1}^{n} f_k(x) \right| < \varepsilon \quad \text{for all } x \in E \text{ and all } m, n \text{ with } N \leq m < n$$

whenever $\varepsilon > 0$ is given. Calling $S = \sum_{n=1}^{\infty} M_n$, we choose N large enough to make $\left| S - \sum_{n=1}^{N} M_n \right| < \varepsilon$. By hypothesis,

$$\left| \sum_{k=m+1}^{n} f_k(x) \right| \leq \sum_{k=m+1}^{n} |f_k(x)| \leq \sum_{k=m+1}^{n} M_k \quad \text{for all } x \in E.$$

Since $\sum_{n=1}^{\infty} M_n$ can't have any negative terms, we see that

$$\sum_{k=m+1}^{n} M_k \leq S - \sum_{k=1}^{N} M_k < \varepsilon \text{ for } N \leq m < n.$$

That completes the proof.   ■

The notion of uniform convergence is an extremely useful one, and we'll see other theorems in which it appears as a hypothesis. But it's a stronger condition than we may really need when we want to establish that a function defined as a limit is continuous at all points in its domain. To prove continuity of a function f on an interval I, we prove continuity at each point in I. For that we only need to consider what happens in a small subinterval about each point. Consequently, we may be able to establish continuity on an interval by using uniform convergence on subintervals instead of on the whole interval. For example, if $I = (-1, 1)$ and $f_n(x) = x^n$, then $f_n(x) \to 0$ at each $x \in I$. But the convergence is not uniform on I; for $\varepsilon > 0$ we see $f_n(x) \geq \varepsilon$ when $\sqrt[n]{\varepsilon} \leq x < 1$. Thus for $0 < \varepsilon < 1$ it's not possible to choose an n such that $|f_n(x)| < \varepsilon$ for all $x \in I$. However, for any $x_0 \in (-1, 1)$, we can find a, b with $x \in (a, b)$ and $[a, b] \subset (-1, 1)$. Uniform convergence is easy to establish on the subinterval $[a, b]$. Note $|f_n(x)|$ is bounded by the larger of $|a^n|$ and $|b^n|$ when $x \in [a, b]$ and we know both $a^n \to 0$ and $b^n \to 0$.

EXERCISES

3. According to the formula for the sum of an infinite geometric series,

$$\sum_{n=0}^{\infty} x^n = \frac{1}{1-x} \quad \text{for } x \in (-1, 1).$$

Show that the sequence of partial sums for this series does not converge uniformly on $(-1, 1)$, but that it does converge uniformly on each interval $[-a, a]$ with $0 < a < 1$.

4. For f a given continuous function on $\mathbf{R}$, let $\{f_n\}_{n=1}^\infty$ be the sequence of continuous functions on $\mathbf{R}$ defined by

$$f_n(x) = n \left[f\left(x + \frac{1}{n}\right) - f(x) \right].$$

Show that $\{f_n\}_{n=1}^\infty$ converges pointwise if f is differentiable on $\mathbf{R}$ and that the convergence is uniform if f' is also differentiable on $\mathbf{R}$ with f'' bounded, but is not uniform on any interval where f' is discontinuous.

5. In Chapter 6, we showed that $(1 + x/n)^n \to e^x$ for all $x \in \mathbf{R}$. Show that the convergence is not uniform. (An easy way is to look at $|(1 + x/n)^n - e^x|$ with $x = n$.)

3 INTEGRALS AND DERIVATIVES

Now that we know sequences of continuous functions can be used to define new continuous functions, we need to try to answer the usual questions we ask about any function we encounter in calculus: Is it differentiable? Is there a formula for its derivative? Is it integrable? Is there a formula for its integral? In this case, we have an obvious place to look for answers: the sequence we use to define the function.

While we often think of differentiation as simpler than integration, the questions about integration are much easier to answer. It is surprisingly difficult to find conditions guaranteeing that term-by-term differentiation of a sequence of functions will produce a sequence converging to the derivative. We'll take a backdoor approach to that problem and tackle the question of integration first. Then we'll use the fundamental theorem of calculus to convert questions of differentiability to questions of integrating derivatives.

THEOREM 3.1: *Let $\{f_n\}_{n=1}^\infty$ be a sequence of Riemann integrable functions on $[a, b]$. If $f_n \to f$ uniformly, then f is also Riemann integrable over $[a, b]$, with*

$$\int_a^b f(x)\, dx = \lim_{n \to \infty} \int_a^b f_n(x)\, dx.$$

□ *Proof:* We'll use Theorem 2.2 in Chapter 5 to prove that f is Riemann integrable, then show that $\int_a^b f(x)\,dx$ is the limit of the sequence of integrals.

To apply Theorem 2.2, we must show that whenever $\varepsilon > 0$ is given, there is a partition $\mathcal{P}$ such that the set $\mathcal{R}(\mathcal{P})$ of Riemann sums for f has diameter less than ε. Let's call $\mathcal{R}_n(\mathcal{P})$ the corresponding set of Riemann sums for f_n. We begin by choosing an n for which

$$|f(x) - f_n(x)| < \frac{\varepsilon}{4(b-a)} \quad \text{for all } x \in [a,b];$$

uniform convergence allows this. If $\mathcal{P} = \{x_0, x_1, \ldots, x_m\}$ is any partition and $\{x_1^*, \ldots, x_m^*\}$ is any set of sampling points for $\mathcal{P}$, the corresponding elements of $\mathcal{R}(\mathcal{P})$ and $\mathcal{R}_n(\mathcal{P})$ satisfy

$$\left| \sum_{k=1}^m f(x_k^*)\Delta x_k - \sum_{k=1}^m f_n(x_k^*)\Delta x_k \right| \le \sum_{k=1}^m |f(x_k^*) - f_n(x_k^*)|\Delta x_k$$

$$< \sum_{k=1}^m \frac{\varepsilon}{4(b-a)}\Delta x_k = \frac{1}{4}\varepsilon.$$

Consequently, we must have

$$\operatorname{diam}\mathcal{R}(\mathcal{P}) \le \operatorname{diam}\mathcal{R}_n(\mathcal{P}) + \frac{1}{2}\varepsilon,$$

and that is true for every partition $\mathcal{P}$ of $[a,b]$. By hypothesis, f_n is Riemann integrable, and so there must be a partition $\mathcal{P}$ with $\operatorname{diam}\mathcal{R}_n(\mathcal{P}) < \frac{1}{2}\varepsilon$; that is the partition we need.

To conclude the proof, we need to show that for any $\varepsilon > 0$ there is an N such that

$$\left| \int_a^b f_n(x)\,dx - \int_a^b f(x)\,dx \right| < \varepsilon \quad \text{for all } n \ge N.$$

Uniform convergence makes that easy; we choose our N to make

$$|f_n(x) - f(x)| < \frac{\varepsilon}{2(b-a)}$$

for all $n \ge N$ and all $x \in [a,b]$. ∎

With the help of Theorem 3.1, it's easy to develop conditions that will justify term-by-term differentiation of a convergent sequence of functions. The theorem below gives them.

THEOREM 3.2: *Let I be an open interval, and suppose that $\{f_n\}_{n=1}^{\infty}$ is a sequence of differentiable functions on I, with each derivative f_n' continuous on I. Suppose further that the sequence $\{f_n'\}_{n=1}^{\infty}$ converges uniformly on each closed subinterval of I, and that there is a point $a \in I$ at which $\{f_n(a)\}_{n=1}^{\infty}$ converges. Then the sequence $\{f_n\}_{n=1}^{\infty}$ converges pointwise on I to a differentiable function f, with*

$$f'(x) = \lim_{n \to \infty} f_n'(x) \quad \text{for each } x \in I.$$

☐ *Proof:* First we assign names to some of the limits that we know exist. In particular, we define

$$f(a) = \lim_{n \to \infty} f_n(a).$$

Since $\{f_n'(x)\}_{n=1}^{\infty}$ must converge for each $x \in I$, we can also define a function g on I by

$$g(x) = \lim_{n \to \infty} f_n'(x).$$

Given any $x \in I$, we can always find an open interval (c, d) containing x and having $[c, d] \subset I$. Since $f_n' \to g$ uniformly on $[c, d]$, Theorem 2.1 guarantees that g is continuous on $[c, d]$, and in particular, g is continuous at x.

Since g is continuous on I, we can also define a function f on I by the formula

$$f(x) = f(a) + \int_a^x g(t)\, dt.$$

That makes f differentiable on I, with

$$f'(x) = g(x) = \lim_{n \to \infty} f_n'(x) \quad \text{for all } x \in I.$$

To complete the proof, we just need to show that

$$f(x) = \lim_{n \to \infty} f_n(x) \quad \text{for all } x \in I.$$

Our definitions guarantee $f_n(a) \to f(a)$, and when $x \neq a$ we may use

$$f_n(x) = f_n(a) + \int_a^x f_n'(t)\, dt.$$

This gives us

$$\lim_{n\to\infty} f_n\left(x\right) = \lim_{n\to\infty} f_n\left(a\right) + \lim_{n\to\infty} \int_a^x f_n'\left(t\right)\, dt$$

$$= f\left(a\right) + \int_a^x g\left(t\right)\, dt = f\left(x\right)$$

by Theorem 3.1. Note that Theorem 3.1 applies directly to integration over $[a, x]$ when $a < x$, and when $x < a$ we can apply it to integration over $[x, a]$ instead.   ∎

A simple example illustrates the differences between these two theorems very nicely. Consider

$$f_n\left(x\right) = \frac{1}{n}\sin n^2 x, \quad -\infty < x < \infty.$$

Then $f_n \to 0$ uniformly, and the limit is a function that is certainly continuous, integrable, and differentiable. But while

$$\left| \int_a^b f_n\left(x\right)\, dx - \int_a^b 0\, dx \right| = \frac{1}{n^3}\left| \cos n^2 a - \cos n^2 b \right| \to 0,$$

we see that

$$\left| f_n'\left(x\right) - \frac{d}{dx}\left(0\right) \right| = \left| n \cos n^2 x \right|,$$

and this can get very large as $n \to \infty$.

EXERCISES

6. For $f\left(x\right) = \sum_{n=1}^{\infty} f_n\left(x\right)$, find conditions under which

$$\int_a^b f\left(x\right)\, dx = \sum_{n=1}^{\infty} \int_a^b f_n\left(x\right)\, dx,$$

and find conditions under which

$$f'\left(x\right) = \sum_{n=1}^{\infty} f_n'\left(x\right).$$

Note that Theorems 3.1 and 3.2 can be used for this purpose.

7. Apply the first exercise to the geometric series to show that

$$-\log\left(1 - x\right) = \int_0^x \frac{dt}{1-t} = \sum_{n=0}^{\infty} \frac{x^{n+1}}{n+1} \quad \text{for } -1 < x < 1.$$

4 TAYLOR'S THEOREM

In Chapter 4 we learned how to use linear polynomials to approximate differentiable functions and how to use quadratic polynomials to approximate functions having second derivatives. Our goal here is a general result, known as Taylor's theorem, that deals with using polynomials of degree n to approximate a function on an interval I. It's named after the English mathematician Brook Taylor, a younger contemporary of Isaac Newton. As we mentioned in Chapter 4, one version of Taylor's theorem can be derived in much the same way we derived the formula for the error in the quadratic approximation near a point. The version we give here is less difficult to justify.

THEOREM 4.1: *Suppose I is an open interval, f is a function in $C^{n+1}(I)$, and $a \in I$. Then for all $x \in I$ we may write*

$$f(x) = \sum_{k=0}^{n} \frac{1}{k!} f^{(k)}(a)(x-a)^k + \frac{1}{n!} \int_a^x f^{(n+1)}(t)(x-t)^n \, dt.$$

The polynomial

$$P_n(x) = \sum_{k=0}^{n} \frac{1}{k!} f^{(k)}(a)(x-a)^k$$

$$= f(a) + f'(a)(x-a) + \cdots + \frac{1}{n!} f^{(n)}(a)(x-a)^n$$

is called either the nth order **Taylor polynomial** or **Taylor's approximation** for $f(x)$ near $x = a$. It's the one polynomial of degree n that satisfies

$$P_n^{(k)}(a) = f^{(k)}(a) \quad \text{for } k = 0, 1, 2, \dots, n.$$

The last expression in Theorem 4.1 is usually thought of as representing the error when $f(x)$ is approximated by $P_n(x)$. Note that if M is a number such that $\left| f^{(n+1)}(t) \right| \leq M$ for all t in the interval between a and x, then since $(x-t)^n$ doesn't change sign on that interval the error term satisfies

$$\left| \frac{1}{n!} \int_a^x f^{(n+1)}(t)(x-t)^{n-1} \, dt \right| \leq \left| \frac{M}{n!} \int_a^x (x-t)^n \, dt \right|$$

$$= \frac{M}{(n+1)!} \left| x - a \right|^{n+1}.$$

☐ *Proof:* The theorem is a consequence of the identity

$$\frac{d}{dt}\left\{\sum_{k=1}^{n}\frac{1}{k!}f^{(k)}\left(t\right)\left(x-t\right)^{k}\right\} = \frac{1}{n!}f^{(n+1)}\left(t\right)\left(x-t\right)^{n} - f'\left(t\right), \quad (8.1)$$

a formula that's simpler than it looks. The derivative of the sum is the sum of the derivatives, and the derivative of the kth term is

$$\frac{1}{k!}f^{(k+1)}\left(t\right)\left(x-t\right)^{k} - \frac{1}{\left(k-1\right)!}f^{(k)}\left(t\right)\left(x-t\right)^{k-1}.$$

So the terms cancel in pairs when we add the derivatives, and the only two that remain form the right-hand side of (8.1). It's also easy to prove (8.1) by induction.

For x and a in I, we can integrate equation (8.1) since all the functions involved are continuous. Integrating the left-hand side gives

$$\int_{a}^{x}\frac{d}{dt}\left\{\sum_{k=1}^{n}\frac{1}{k!}f^{(k)}\left(t\right)\left(x-t\right)^{k}\right\} dt = \sum_{k=1}^{n}\left[\frac{1}{k!}f^{(k)}\left(t\right)\left(x-t\right)^{k}\right]_{a}^{x}$$

$$= -\sum_{k=1}^{n}\frac{1}{k!}f^{(k)}\left(a\right)\left(x-a\right)^{k},$$

while the right-hand side yields

$$\int_{a}^{x}\left[\frac{1}{n!}f^{(n+1)}\left(t\right)\left(x-t\right)^{n} - f'\left(t\right)\right] dt = \frac{1}{n!}\int_{a}^{x}f^{(n+1)}\left(t\right)\left(x-t\right)^{n} dt$$

$$- \left[f\left(x\right) - f\left(a\right)\right].$$

So rearranging the terms in the integrated equation will complete the proof. ∎

Often we use Taylor's approximation with n and a fixed and treat x as the only variable. But sometimes we take a different point of view. When f is in the class $\mathcal{C}^{\infty}\left(I\right)$, we can form the approximation $P_{n}\left(x\right)$ for every n, which defines a sequence $\left\{P_{n}\left(x\right)\right\}_{n=1}^{\infty}$ of approximations to $f\left(x\right)$. Then we may well ask whether this sequence of functions converges pointwise to f on I, as well as whether we can use the sequence to find integrals or derivatives of f. Sometimes we can't, but the cases when it does work are important enough to form the basis for several important definitions.

DEFINITION 4.1:   When I is an open interval, $a \in I$, and $f \in \mathcal{C}^{\infty}\left(I\right)$, the infinite series

$$\sum_{n=0}^{\infty}\frac{1}{n!}f^{(n)}\left(a\right)\left(x-a\right)^{n}$$

is called the **Taylor series** for $f(x)$ based at a. If there is an $r > 0$ such that the Taylor series for $f(x)$ based at a converges to $f(x)$ for all $x \in (a - r, a + r)$, we say that f is **analytic** at a. We call f an **analytic function** if it is analytic at each point in its domain.

Many of the familiar functions of calculus turn out to be analytic functions, but some interesting ones aren't analytic. There are C^∞ functions whose Taylor series based at a converge only when $x = a$, and there are others whose Taylor series converge but not to $f(x)$. We can be sure that a function is analytic at a point if we can prove there is an interval in which the error term approaches zero as $n \to \infty$. But that may be quite difficult to establish unless there is a simple formula giving $f^{(n)}(x)$ for all n. Let's look at two examples.

The natural exponential function is an analytic function. For $f(x) = e^x$, $f^{(n)}(x) = e^x$ for all n, so the Taylor series based at a is

$$\sum_{n=0}^{\infty} \frac{1}{n!} e^a (x - a)^n .$$

According to Theorem 4.1, for any n we have

$$e^x - \sum_{k=0}^{n} \frac{1}{k!} e^a (x - a)^k = \frac{1}{n!} \int_a^x e^t (x - t)^n \, dt.$$

Since $0 \leq e^t \leq e^{a+r}$ for all $t \in (a - r, a + r)$, when $x \in (a - r, a + r)$ we have

$$\left| \frac{1}{n!} \int_a^x e^t (x - t)^n \, dt \right| \leq \frac{e^{a+r}}{n!} \left| \int_a^x (x - t)^n \, dt \right|$$

$$= \frac{e^{a+r}}{(n + 1)!} |x - a|^{n+1} \leq \frac{e^{a+r} r^{n+1}}{(n + 1)!}.$$

No matter what $r > 0$ we consider,

$$\lim_{n \to \infty} \frac{e^{a+r} r^{n+1}}{(n + 1)!} = 0.$$

When $n \geq N \geq 2r - 1$, we have

$$\frac{e^{a+r} r^{n+1}}{(n + 1)!} = \frac{e^{a+r} r^N}{N!} \cdot \frac{r}{N + 1} \cdot \frac{r}{N + 2} \cdots \frac{r}{n + 1}$$

$$\leq \frac{e^{a+r} r^N}{N!} \left(\frac{1}{2} \right)^{n+1-N} \to 0 \text{ as } n \to \infty.$$

We've proved that for any $x, a \in \mathbf{R}$ we have

$$e^x = \sum_{n=0}^{\infty} \frac{1}{n!} e^a (x - a)^n .$$

The case $a = 0$ is especially useful, because $e^0 = 1$. The formula

$$e^x = \sum_{n=0}^{\infty} \frac{1}{n!} x^n$$

gives a useful way to calculate values for the exponential function when $|x|$ isn't too large.

Another simple example of an analytic function is given by $f(x) = 1/x$; it's analytic at every $a \neq 0$. Using induction, it's easy to show that

$$f^{(n)}(x) = \frac{(-1)^n n!}{x^{n+1}},$$

so the Taylor series for $1/x$ based at a is

$$\sum_{n=0}^{\infty} \frac{(-1)^n}{a^{n+1}} (x - a)^n .$$

According to Theorem 4.1, as long as a and x are both positive or both negative we have

$$\frac{1}{x} - \sum_{k=0}^{n} \frac{(-1)^k}{a^{k+1}} (x - a)^k = \frac{1}{n!} \int_a^x \frac{(-1)^{n+1} (n+1)!}{t^{n+2}} (x - t)^n \, dt$$

$$= -(n+1) \int_a^x \left(1 - \frac{x}{t}\right)^n \frac{dt}{t^2}$$

$$= -\frac{1}{x} \left(1 - \frac{x}{t}\right)^{n+1} \Big|_a^x = \frac{1}{x} \left(1 - \frac{x}{a}\right)^{n+1} .$$

When $|1 - x/a| < 1$, the error term approaches zero as $n \to \infty$, which proves that

$$\frac{1}{x} = \sum_{n=0}^{\infty} \frac{(-1)^n}{a^{n+1}} (x - a)^n \quad \text{for all } x \in (a - |a|, a + |a|).$$

We conclude this section with a caution against reading something into the definition of *analytic* that isn't there. When f is analytic and a is a point in its domain, the domain of f can be quite different from the set of points at which the power series based at a converges to f. Here we will not investigate the relationship between those sets; that's a matter best studied in another branch of mathematics called complex analysis.

EXERCISES

8. Explain why every polynomial function $p(x)$ is analytic on **R**.

9. Show that $\sin x$ and $\cos x$ are analytic on **R** by noting that the error terms are always

$$\pm \frac{1}{n!} \int_a^x (x - t)^n \cos t \, dt \quad \text{or} \quad \pm \frac{1}{n!} \int_a^x (x - t)^n \sin t \, dt,$$

and then proving that the error terms approach 0 as $n \to \infty$.

10. The function f defined by

$$f(x) = \begin{cases} e^{-1/x}, & x > 0 \\ 0, & x \le 0 \end{cases}$$

is the simplest example of a C^∞ function that is not analytic. Of course $f^{(n)}(x)$ is zero for all $x < 0$, and for $x > 0$ it has the form $e^{-1/x} P_n(1/x)$ with P_n a polynomial of degree $2n$. The polynomials can be defined inductively by the rules $P_0(t) = 1$ and

$$P_{n+1}(t) = t^2 \left[P_n(t) - P_n'(t) \right].$$

Complete the argument that $f \in C^\infty$ by showing that

$$f^{(n+1)}(0) = \lim_{x \to 0} \frac{f^{(n)}(x) - f^{(n)}(0)}{x} = 0$$

for all integers $n \ge 0$. Why does it follow that f is not analytic at 0?

5 POWER SERIES

Any series of the form

$$\sum_{n=0}^{\infty} c_n (x - a)^n$$

is called a **power series** as long as each coefficient c_n is independent of x; we call a the **base point** for the series. Obviously a Taylor series based at a will be a power series with base point a. One of our goals in this section is to establish a sort of converse to this idea, giving a simple condition for a power series to be the Taylor series of an analytic function. That will help us understand how calculus operations can be performed on analytic functions by manipulating their Taylor series.

One of the simplest power series turns out to be the most important one to understand: the geometric series $\sum_{n=0}^{\infty} x^n$. Just about everything we can ever learn about general power series is based on what we understand about this one case, and we understand a lot because we have a convenient formula for the partial sums:

$$\sum_{k=0}^{n} x^k = \frac{1 - x^{n+1}}{1 - x} \quad \text{when } x \neq 1.$$

Since $\left| x^{n+1} \right| \to 0$ if $|x| < 1$ and $\left| x^{n+1} \right| \to \infty$ if $|x| > 1$, we see that the geometric series converges for $|x| < 1$ and diverges for $|x| > 1$. The cases $x = 1$ and $x = -1$ should be handled separately. When $x = 1$ we must abandon our formula for the partial sums and use common sense instead:

$$\sum_{k=0}^{n} x^k = n \quad \text{when } x = 1.$$

So at $x = 1$ the series diverges to ∞. At $x = -1$ the series also diverges, but in a different way; the partial sums alternate between 0 and 1 instead of either growing or approaching a limiting value. We conclude that

$$\sum_{n=0}^{\infty} x^n = \frac{1}{1 - x} \quad \text{for } |x| < 1$$

and the series diverges for $|x| \geq 1$.

Next we ask whether term-by-term differentiation of the geometric series produces a new series that converges to the derivative of $1/(1 - x)$. For $x \neq 1$, we can differentiate our formula for the partial sums to obtain

$$\sum_{k=1}^{n} k x^{k-1} = \frac{d}{dx} \left[\sum_{k=0}^{n} x^k \right] = \frac{d}{dx} \left[\left(1 - x^{n+1} \right) \left(1 - x \right)^{-1} \right]$$

$$= -\frac{(n + 1) x^n}{1 - x} + \frac{1 - x^{n+1}}{(1 - x)^2}. \tag{8.2}$$

Consequently, when both $x^{n+1} \to 0$ and $(n+1)\,x^n \to 0$ as $n \to \infty$, we obtain

$$\sum_{n=1}^{\infty} nx^{n-1} = \lim_{n \to \infty} \sum_{k=1}^{n} kx^{k-1} = \lim_{n \to \infty} \left[-\frac{(n+1)\,x^n}{1-x} + \frac{1 - x^{n+1}}{(1-x)^2} \right]$$

$$= \frac{1}{(1-x)^2},$$

which is the derivative of $1/(1-x)$. We know $x^{n+1} \to 0$ for $|x| < 1$; we'll show that $(n+1)\,x^n \to 0$ for the same values of x.

All we really need to consider is the case $0 < x < 1$; the case $x = 0$ presents no challenge and for $x < 0$ we can use $|(n+1)\,x^n| = (n+1)\,|x|^n$. For $0 < x < 1$, we can drop the terms in (8.2) with minus signs to obtain

$$\sum_{k=1}^{n} kx^{k-1} < \frac{1}{(1-x)^2}$$

for all n. Regarding x as fixed, we define

$$M = \sup \left\{ \sum_{k=1}^{n} kx^{k-1} : n \in \mathbf{N} \right\}.$$

Given $\varepsilon > 0$, there must be an N with

$$\sum_{k=1}^{N} kx^{k-1} > M - \varepsilon.$$

Since the terms in the sum are all positive, we have $(n+1)\,x^n < \varepsilon$ for all $n \geq N$, proving that $(n+1)\,x^n \to 0$ for $|x| < 1$.

Now we're ready to tackle more general power series. The theorem below is an important step.

THEOREM 5.1: *At any x where the power series $\sum_{n=0}^{\infty} c_n\,(x-a)^n$ converges, the set of numbers $\{ c_n\,(x-a)^n : n \in \mathbf{N} \}$ is a bounded set. Conversely, if r is a positive number such that $\{ c_n r^n : n \in \mathbf{N} \}$ is a bounded set, then the power series $\sum_{n=0}^{\infty} c_n\,(x-a)^n$ converges pointwise to a differentiable function $f(x)$ on $(a-r, a+r)$, and the differentiated series $\sum_{n=1}^{\infty} nc_n\,(x-a)^{n-1}$ converges to $f'(x)$ on the same interval. The convergence of both series is uniform on every closed subinterval of $(a-r, a+r)$.*

□ *Proof:* The power series converges when the sequence of partial sums $\left\{\sum_{k=0}^{n} c_k (x-a)^k\right\}_{n=1}^{\infty}$ converges, and the terms in any convergent sequence form a bounded set. Hence there must be a number M (depending on x) such that all the partial sums are in $[-M, M]$. Therefore

$$|c_n (x-a)^n| \leq 2M$$

since each term is the difference of two consecutive partial sums.

For the converse, we'll prove the last part first and then use Theorem 3.2 to justify term-by-term differentiation of the series. Given a closed subinterval $I \subset (a-r, a+r)$, we prove that the differentiated series converges uniformly on I by using the Weierstrass M-test, as given in Theorem 2.2. To use the M-test, we need a number r_0 smaller than r but large enough that $I \subset [a-r_0, a+r_0]$. Such a number exists since I is a closed subinterval of $(a-r, a+r)$. Then for all $x \in I$ we have

$$\left| kc_k (x-a)^{k-1} \right| \leq k |c_k| r_0^{k-1} = k \left| c_k r^{k-1} \right| (r_0/r)^{k-1}.$$

We've assumed that there is a number M with $\left| c_k r^k \right| \leq M$ for all k, and so

$$\left| kc_k (x-a)^{k-1} \right| \leq kMr^{-1} (r_0/r)^{k-1} \quad \text{for all } x \in I.$$

Note that $\sum_{k=1}^{\infty} k (r_0/r)^{k-1}$ converges since $0 < r_0/r < 1$, and the series remains convergent when we multiply by Mr^{-1}. So by the M-test, the differentiated series converges uniformly on I. We can use the same argument to prove that the original series converges uniformly on I, except our bound is

$$\left| c_k (x-a)^k \right| \leq M (r_0/r)^k.$$

Of course, Theorem 3.2 applies to sequences, not series, so we must apply it to the sequence of partial sums instead of the terms in the series. Obviously

$$\frac{d}{dx}\left[\sum_{k=0}^{n} c_k (x-a)^k \right] = \sum_{k=1}^{n} kc_k (x-a)^{k-1},$$

so the limit of the sequence of derivatives is indeed the sum of the differentiated series. That completes the proof.   ∎

Under the hypotheses of Theorem 5.1, we can say a great deal more about the function defined by the sum of the power series. The theorem below is a significant step in that direction.

THEOREM 5.2: *If the power series $\sum_{n=0}^{\infty} c_n (x - a)^n$ converges pointwise on $(a - r, a + r)$ to a function f, then f is in $C^{\infty} (a - r, a + r)$, and derivatives of all orders may be found by using term-by-term differentiation of the power series.*

☐ *Proof:* All we really need to do is show that the hypotheses guarantee that the differentiated series

$$\sum_{n=1}^{\infty} n c_n (x - a)^{n-1}$$

converges to $f'(x)$ on $(a - r, a + r)$. Then the theorem will follow by induction; every time we take a derivative we get a new power series that converges on the same interval, so we can always differentiate one more time.

Given $x \in (a - r, a + r)$, we can prove that

$$f'(x) = \sum_{n=1}^{\infty} n c_n (x - a)^{n-1}$$

by using Theorem 5.1. Here, however, we don't know that $\{c_n r^n\}_{n=0}^{\infty}$ is a bounded sequence, so we can't use the theorem directly. Regarding x as fixed for the moment, we can pick r_0 with $|x - a| < r_0 < r$ and apply the theorem on $(a - r_0, a + r_0)$ instead. Note that $\sum_{n=0}^{\infty} c_n r_0^n$ converges to $f(a + r_0)$ by hypothesis, and the first part of Theorem 5.1 shows that $\{c_n r_0^n\}_{n=1}^{\infty}$ must be a bounded sequence. ∎

COROLLARY 5.1: *A function defined on an open interval by means of a convergent power series is analytic at the base point, and the power series is the Taylor series for the function at that point.*

To prove the corollary, we only need to recognize that we do indeed have power series expressions for derivatives of all orders since evaluating any power series at the base point picks out the constant term.

The corollary can be quite useful because there are quite a few ways to produce convergent power series. For example, replacing x by x^2 in the power series for e^x based at 0 shows that

$$f(x) = e^{x^2} = \sum_{n=0}^{\infty} \frac{1}{n!} x^{2n}$$

defines a function that is analytic at 0, with

$$f^{(k)}(0) = \begin{cases} 0, & k \text{ odd} \\ (2n)!/n!, & k = 2n. \end{cases}$$

On the other hand, using elementary calculus formulas to discover a general formula for the kth derivative of e^{x^2} is difficult because the number of terms involved keeps growing.

We'll end this section by pointing out that Theorem 5.1 can also be used to justify term-by-term integration of a power series since each partial sum is continuous and the sequence of partial sums converges uniformly on each closed subinterval of $(a - r, a + r)$.

EXERCISES

11. Prove that for any power series $\sum_{n=0}^{\infty} c_n (x - a)^n$, the set of all x for which it converges is an interval.

12. Prove that the power series $\sum_{n=0}^{\infty} c_n (x - a)^n$ converges for all x in the interval $(a - r, a + r)$ if there is a number N such that

$$r |c_n|^{1/n} \leq 1 \quad \text{for all } n \geq N.$$

13. Use the example of the geometric series to show why convergence of a power series at each point in $(a - r, a + r)$ does not justify term-by-term integration over $[a - r, a + r]$.

6   TOPICS FOR FURTHER STUDY

When a power series $\sum_{n=0}^{\infty} c_n (x - a)^n$ converges for some $x \neq a$ but not for all x, there is a number $R > 0$ such that the series converges on $(a - R, a + R)$ but diverges at all points outside $[a - R, a + R]$. We call R the **radius of convergence**. There are formulas for computing it from the coefficients, but they involve expressions that we have no effective means to evaluate unless the coefficients conform to some regular pattern.

Extensive studies of analytic functions usually deal with functions defined on a region of the complex plane rather than just along the real line. That allows for the development of more effective methods than we could use here. In particular, there are easy ways to show that most elementary functions are analytic and that various combinations of analytic functions are also analytic. This is part of the subject matter of the theory of functions of a complex variable, the subject of many mathematics books. Gauss, Riemann, Weierstrass, and Cauchy produced major contributions to this branch of mathematics, and some of the world's best mathematicians are still trying to settle a question first raised by Riemann about the location of the zeroes of a particular analytic function.

Power series are not the only infinite series of functions that have been studied extensively; the French mathematician Joseph Fourier introduced series of the form

$$a_0 + \sum_{n=1}^{\infty} (a_n \cos n\theta + b_n \sin n\theta)$$

to study problems of heat flow. The study of such series, now called Fourier series, has been a major part of twentieth century mathematics and has developed into a discipline now called **harmonic analysis**.

Fourier series can define functions with very strange properties, such as the example we gave of a function that is everywhere continuous but nowhere monotonic. Prior to the time of Fourier, the sum of an infinite series of functions was thought to share any properties common to all the terms, such as continuity and differentiability, but Fourier claimed otherwise. The controversy could not be resolved until precise definitions were given for all the concepts involved. At this stage it was clear that intuition and reasoning from simple examples did not provide a sound basis for calculus.

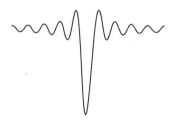

IX

ADDITIONAL COMPUTATIONAL METHODS

n this final chapter we study four independent topics that are often introduced in first-year calculus courses. One common thread linking them is that each is designed to produce a numerical value by avoiding a direct calculation of the desired quantity.

I L'HÔPITAL'S RULE

In calculus we often consider the limit of a quotient at a point where both the numerator and denominator vanish. Calculating the value of a derivative by using its definition as a limit is one such problem, for example. Of course, we seldom calculate a derivative by working directly with the limit; we usually have more efficient ways to find derivatives. The ease of finding derivatives suggests investigating whether we can use them to simplify the calculation of additional limits of this type. To start our investigation, let's suppose that f_1 and f_2 are numerical functions defined on an interval I, that $a \in I$ with $f_2(x) \neq f_2(a)$ for any $x \in I$ except for a itself, and that we need to find

$$\lim_{x \to a} \frac{f_1(x) - f_1(a)}{f_2(x) - f_2(a)}.$$

If we assume that f_1 and f_2 are differentiable at a, then we know there are numerical functions g_1 and g_2 on I that are continuous at a and satisfy

$$f_1(x) = f_1(a) + (x - a) g_1(x)$$

and $\qquad f_2(x) = f_2(a) + (x - a) g_2(x).$

Thus

$$\frac{f_1(x) - f_1(a)}{f_2(x) - f_2(a)} = \frac{g_1(x)}{g_2(x)} \quad \text{for all } x \neq a \text{ in } I. \tag{9.1}$$

If $g_2(a) \neq 0$, then by continuity of g_1 and g_2 at a we have

$$\lim_{x \to a} \frac{f_1(x) - f_1(a)}{f_2(x) - f_2(a)} = \frac{g_1(a)}{g_2(a)} = \frac{f_1'(a)}{f_2'(a)},$$

so our limit is simply the quotient of the values of the derivatives.

However, in the course of evaluating such limits, we may find that

$$f_1'(a) = f_2'(a) = 0,$$

and then equation (9.1) isn't much help. For example, in evaluating

$$\lim_{x \to 0} \frac{1 - \cos x}{\exp(x^2) - 1}$$

we see that the derivatives of the numerator and denominator are $\sin x$ and $2x \exp(x^2)$; both vanish at $x = 0$. For $x \neq 0$ we have $g_1(x) = (1 - \cos x)/x$ and $g_2(x) = [\exp(x^2) - 1]/x$, and there's no reason to use $g_1(x)/g_2(x)$ in place of the original quotient. However, once we remember that $(\sin x)/x \to 1$, we have no trouble recognizing that

$$\lim_{x \to 0} \frac{f_1'(x)}{f_2'(x)} = \lim_{x \to 0} \frac{\sin x}{2x \exp(x^2)} = \frac{1}{2}.$$

But is that the same as the limit of $g_1(x)/g_2(x)$? The answer is yes, as a consequence of **L'Hôpital's rule,** a useful tool for analyzing limits of quotients in many cases such as this one.

This rule appeared in what is thought to be the world's first calculus textbook, published in 1696 by G. F. A. de L'Hôpital, a French marquis.

It's generally believed that he did not discover this formula, but that it was discovered by John Bernoulli, a member of a remarkable family of mathematicians and scientists. L'Hôpital had hired Bernoulli to teach him mathematics, and the contract included the rights to use Bernoulli's discoveries. After L'Hôpital's death, Bernoulli claimed the formula was his, but L'Hôpital's name remains.

The use of L'Hôpital's rule is not restricted to finding limits exactly of the form we assumed in the first paragraph. More generally we assume that we are looking for a limit of a quotient, and the limit can't be found by taking the limits of the numerator and denominator separately. In addition to finding the limit of $f(x)/g(x)$ at a point where both f and g vanish, it includes limits at points where both f and g approach ∞ and limits at ∞ as well as at points in **R**. All the different cases can be analyzed with a few variants of a single argument using Cauchy's extension of the mean value theorem, Theorem 3.3 in Chapter 4.

To set up the argument, we'll assume that f and g are differentiable on an open interval I, with neither g nor g' ever vanishing on I, and that the limit we need is determined by the values of $f(x)/g(x)$ with $x \in I$. This framework applies directly to one-sided limits and limits at ∞ or $-\infty$; two-sided limits can be analyzed by considering left-hand and right-hand limits separately. With our assumptions, Theorem 3.3 tells us that for any $x, y \in I$ with $x \neq y$, there must be a point ξ between x and y with

$$\frac{f(x) - f(y)}{g(x) - g(y)} = \frac{f'(\xi)}{g'(\xi)}.$$

That lets us write

$$\frac{f(x)}{g(x)} = \frac{f(y)}{g(x)} + \frac{g(x) - g(y)}{g(x)} \cdot \frac{f(x) - f(y)}{g(x) - g(y)}$$

$$= \frac{f(y)}{g(x)} + \left[1 - \frac{g(y)}{g(x)}\right] \frac{f'(\xi)}{g'(\xi)}. \tag{9.2}$$

This formula allows us to approximate values of f/g by values of f'/g' when $|f(y)|$ and $|g(y)|$ are small in comparison to $|g(x)|$. Before we start considering cases, let's give a formal statement of what we'll prove.

THEOREM 1.1: *Suppose that f and g are differentiable functions on (a, ∞), with neither g nor g' ever vanishing there. Suppose further that either*

$$\lim_{x \to \infty} |g(x)| = \infty \quad or \quad \lim_{x \to \infty} f(x) = \lim_{x \to \infty} g(x) = 0.$$

Then if $f'(x)/g'(x)$ has a limit at ∞ (finite or infinite), $f(x)/g(x)$ has exactly the same limit at ∞.

Similar results hold for limits at $-\infty$ and for left-hand or right-hand limits at a finite point when the interval (a, ∞) is replaced by an appropriate open interval.

☐ *Proof:*   We'll give the proof only for limits at ∞; our argument is easily adapted to all the other possibilities.

First we suppose that f'/g' has the finite limit L. Then we need to show that for every $\varepsilon > 0$, there is a b such that

$$\left| \frac{f(x)}{g(x)} - L \right| < \varepsilon \quad \text{for all } x > b.$$

The key to this inequality is the representation for $f(x)/g(x)$ in equation (9.2); our hypotheses guarantee it is valid for each $x, y \in (a, \infty)$ with $x \neq y$. The idea is to make $|f(y)/g(x)|$ be smaller than $\varepsilon/2$ and also to make

$$\left| \left[1 - \frac{g(y)}{g(x)} \right] \frac{f'(\xi)}{g'(\xi)} - L \right| < \frac{\varepsilon}{2}.$$

We use our rules for approximate arithmetic to do this; there is a $\delta > 0$ such that every u and v with $|u - 1| < \delta$ and $|v - L| < \delta$ satisfy $|uv - L| < \varepsilon/2$. We'll choose our b so that for all $x > b$, there is a point $y \neq x$ in (a, ∞) such that

$$\left| \frac{f(y)}{g(x)} \right| < \frac{\varepsilon}{2} \quad \text{and} \quad \left| \frac{g(y)}{g(x)} \right| < \delta, \tag{9.3}$$

with every ξ between x and y satisfying

$$\left| \frac{f'(\xi)}{g'(\xi)} - L \right| < \delta. \tag{9.4}$$

When f and g both approach 0, we can choose $b \geq a$ such that (9.4) is satisfied for all $\xi > b$. Then for each $x > b$,

$$\lim_{y \to \infty} \frac{f(y)}{g(x)} = \lim_{y \to \infty} \frac{g(y)}{g(x)} = 0,$$

so any sufficiently large y will satisfy (9.3), and for $y > x > b$ every value of ξ between x and y will satisfy (9.4).

The case when $|g(x)| \to \infty$ requires a different treatment. We first fix $y > a$ such that (9.4) is satisfied by all $\xi > y$. Then since

$$\lim_{x \to \infty} \frac{f(y)}{g(x)} = \lim_{x \to \infty} \frac{g(y)}{g(x)} = 0,$$

there is a $b > y$ such that every $x > b$ will satisfy (9.3) and will make every value of ξ between x and y satisfy (9.4) as well.

When f'/g' has an infinite limit, we have a different sort of inequality to satisfy. Let's assume $f'(x)/g'(x) \to \infty$; we can treat a limit of $-\infty$ by considering $-f/g$. Then our goal is to prove that for every $M > 0$ there is a $b \geq a$ such that

$$\frac{f(x)}{g(x)} > M \quad \text{for all} \quad x > b.$$

We still use equation (9.2), but our strategy is to make

$$\left| \frac{f(y)}{g(x)} \right| < 1, \quad \left| \frac{g(y)}{g(x)} \right| < \frac{1}{2}, \text{ and } \frac{f'(\xi)}{g'(\xi)} > 2(M+1).$$

The rest of the argument follows the previous pattern for a finite limit. When f and g approach 0, we use the last inequality to choose b, and then for $x > b$ we can find $y > x$ to satisfy the first two inequalities. When $|g|$ approaches infinity, we use the last inequality to choose y, and then choose $b > y$ so that the first two inequalities are satisfied for all $x > b$. The reader is encouraged to write out the details. ∎

EXERCISES

1. Assuming that

$$\lim_{x \to a} \frac{f'(x)}{g'(x)} = L,$$

what further assumptions are needed to guarantee that

$$\lim_{x \to a} \frac{f(x)}{g(x)} = L?$$

2. Write out a proof of L'Hôpital's rule for the case

$$\lim_{x \to a^+} \frac{f'(x)}{g'(x)} = L.$$

3. For $f(x) = x + \sin x$ and $g(x) = x + 1$, show that

$$\lim_{x \to \infty} \frac{f(x)}{g(x)} = 1$$

and that $f'(x)/g'(x)$ has no limit as $x \to \infty$. Why doesn't this contradict L'Hôpital's rule?

2 NEWTON'S METHOD

From time to time we need extremely accurate numerical values for solutions to equations that we can't solve algebraically. We may be able to use the intermediate value theorem to prove that there is a solution in a particular interval, and we can then use the bisection method to locate a solution to within an arbitrarily high degree of accuracy. But that may take more of an effort than we care to expend. For example, let's consider the equation

$$e^x = 3x,$$

which certainly has a root in $(0, 1)$. Consequently, performing n bisections will reduce the problem to looking in a subinterval of length 2^{-n}. To solve the equation with 12-decimal-place accuracy, we'll need to perform about 40 bisections, and each one requires a new evaluation of e^x. Now imagine that the only available way to compute e^x is to do the calculations by hand; it's obvious that a more efficient way of solving the equation is needed. Fortunately, there are methods that offer the possibility of finding rapid improvements to approximate solutions. One of the best of these is also one of the simplest and one of the oldest and is generally known as **Newton's method**. The availability of greater computing power has made it more important, not less, because the advantages of using Newton's method become more pronounced as higher levels of accuracy are sought. It's now a built-in program on many scientific calculators. But there are significant limitations on its usefulness, and to understand them we need to examine how the method works.

We begin by writing the equation to be solved in the form

$$f(x) = 0,$$

so the solutions are x-intercepts of the graph of f. For the bisection method we require only that f be continuous on an open interval containing the root, but for Newton's method we need f to be differentiable. Let's say that r is a number such that $f(r) = 0$, and that we only know r approximately. If our initial approximation to r is a, we form a new approximation by linearizing $f(x)$ near $x = a$ and then finding where the graph of the linearization crosses the x-axis. That is, we solve

$$f(a) + f'(a)(x - a) = 0.$$

Assuming $f'(a) \neq 0$, the solution is

$$x = a - \frac{f(a)}{f'(a)}.$$

We use this formula to produce a sequence of approximations. Starting with an initial approximation x_1, we define $\{x_n\}_{n=1}^{\infty}$ recursively by calling

$$x_{n+1} = x_n - \frac{f(x_n)}{f'(x_n)} \quad \text{for } f'(x_n) \neq 0. \tag{9.5}$$

Of course, we can halt the process successfully if we find x_n with $f(x_n) = 0$. In practice, we stop when we can no longer recognize a significant difference between 0 and $f(x_n)$. Unfortunately, if we ever encounter x_n with $f'(x_n) = 0$, the only thing we can learn from the experience is not to try the same value of x_1 (or any of $x_2, \ldots, x_{n-1}$ either) as our initial approximation.

Here's how Newton's method works for our equation

$$e^x = 3x,$$

which we rewrite as

$$f(x) = e^x - 3x = 0.$$

Our formula becomes

$$x_{n+1} = x_n - \frac{\exp(x_n) - 3x_n}{\exp(x_n) - 3}.$$

Starting with $x_1 = 0$ and using a calculator that displays 10 digits, we find

$$x_2 = 0.5$$
$$x_3 = 0.610059655$$
$$x_4 = 0.618996780$$
$$x_5 = 0.619061283$$
$$x_6 = 0.619061287,$$

with no further changes recorded.

To explain why Newton's method works so well, we analyze the error in linearizations of f. Theorem 4.1 in Chapter 4 gives us a convenient representation for this error. When f' is differentiable on an interval (α, β) containing both r and the approximation x_n, we can write

$$0 = f(r) = f(x_n) + f'(x_n)(r - x_n) + \frac{1}{2}f''(\xi_n)(r - x_n)^2$$

for some ξ_n between r and x_n. As long as $f'(x_n) \neq 0$,

$$r - x_n = -\frac{f(x_n)}{f'(x_n)} - \frac{f''(\xi_n)}{2f'(x_n)}(r - x_n)^2.$$

Combining this equation with our recursion formula (9.5), we see that

$$r - x_{n+1} = -\frac{f''(\xi_n)}{2f'(x_n)}(r - x_n)^2. \tag{9.6}$$

Now we can pin down some assumptions that will guarantee the success of Newton's method. We're searching for a solution of $f(x) = 0$; we assume that there is a solution $r \in (\alpha, \beta)$ and that there are constants A and B such that

$$\left|f'(x)\right| \geq A > 0 \quad \text{and} \quad \left|f''(x)\right| \leq B \quad \text{for all } x \in (\alpha, \beta).$$

Then if $x_n \in (\alpha, \beta)$, we know that x_{n+1} can be defined and that

$$|r - x_{n+1}| \leq \frac{B}{2A}(r - x_n)^2.$$

There must be an $\varepsilon > 0$ such that

$$(r - \varepsilon, r + \varepsilon) \subset (\alpha, \beta) \quad \text{and} \quad B\varepsilon < A,$$

and then for $0 < |r - x_n| < \varepsilon$ we have

$$|r - x_{n+1}| < \frac{B}{2A}\varepsilon|r - x_n| < \frac{1}{2}|r - x_n|.$$

Consequently, if our initial approximation is within ε of the solution r, then the successive errors will decrease to zero, with the error in the next approximation roughly proportional to the square of the error in the current one. That explains why the accuracy improves so markedly as the approximations get closer to the solution. For the example we worked out, the error in using x_5 was about $4 \cdot 10^{-9}$. If enough decimal places could be used, the error with x_6 would be about $2 \cdot 10^{-17}$ and the error with x_7 would be about $4 \cdot 10^{-34}$.

Newton's method isn't very good for finding repeated roots of equations. When $f(r) = f'(r) = 0$ the problem of finding r becomes quite difficult, no matter what method is used. The bisection method may fail completely; for example, there may well be no x with $f(x) < 0$.

EXERCISES

4. The greatest solution of $x^3 - 4x^2 + 4x - 1 = 0$ is $r = \frac{1}{2}(3 + \sqrt{5})$. Find an $\varepsilon > 0$ such that when we use Newton's method to solve this equation, taking $x_1 \in (r - \varepsilon, r + \varepsilon)$ guarantees that $|x_2 - r| < \frac{1}{2}|x_1 - r|$.

5. What happens when we try to use Newton's method to solve $x^{1/3} = 0$ and start with $x_1 \neq 0$?

6. For the equation $x^3 - 4x^2 + 4x = 0$, show that using Newton's method with $x_n > 2$ always leads to $x_{n+1} - 2 > \frac{1}{2}(x_n - 2)$.

3 SIMPSON'S RULE

Although it is theoretically possible to obtain arbitrarily accurate approximations to integrals by using Riemann sums for partitions with small enough mesh size, in practice this doesn't work. If the mesh of the partition is decreased, the number of calculations required will increase, becoming a source of additional error as well as aggravation. Each calculation performed in terms of decimals can introduce a small round-off error, and the cumulative effect of many small errors may not be small at all. Consequently, mathematicians are always looking for methods to obtain more accuracy with fewer calculations; this is the basic problem of **numerical analysis**. One of the early successes in this area is **Simpson's rule** for approximating integrals. The formula is named for the English mathematician Thomas Simpson, who published it in the eighteenth century.

Calculus books often explain Simpson's rule in terms of parabolic arcs fitted to points on the graph of the function to be integrated. However, this explanation doesn't lead to quantitative expressions for the error in Simpson's rule, but only to a general feeling that the approximation ought to work pretty well. We'll find an analytic expression for the error by deriving an identity for the integral of a numerical function f with four continuous derivatives. The process is similar to our derivation of Taylor's formula in Chapter 8.

We first note that if g is any fourth-degree polynomial with leading coefficient $1/24$, then $g^{(4)}(x) = 1$ and

$$\frac{d}{dx}\left\{\sum_{k=0}^{3}(-1)^k \, g^{(3-k)}(x) \, f^{(k)}(x)\right\} = f(x) - g(x) \, f^{(4)}(x).$$

This lets us use the fundamental theorem of calculus to write

$$\int_a^b f(x) \, dx = \sum_{k=0}^{3}(-1)^k \left[g^{(3-k)}(b) \, f^{(k)}(b) - g^{(3-k)}(a) \, f^{(k)}(a)\right]$$
$$+ \int_a^b g(x) \, f^{(4)}(x) \, dx.$$

To reduce the sum, we make g and its first derivative vanish at both a and b; those conditions determine the remaining coefficients in g. A convenient way to do this is to call $h = b - a$. Then $g(x) = s_h(x - a)$, with s_h defined by

$$s_h(x) = \frac{1}{24}x^2(x-h)^2 \, ;$$

note $g(a) = s_h(0)$ and $g(b) = s_h(h)$. We calculate

$$s_h^{(1)}(x) = \frac{1}{12}x(x-h)(2x-h),$$

$$s_h^{(2)}(x) = \frac{1}{12}\left[(2x-h)^2 + 2x(x-h)\right],$$

$$s_h^{(3)}(x) = \frac{1}{2}(2x-h).$$

So g and $g^{(1)}$ vanish at both a and b, and we also find that

$$g^{(2)}(a) = g^{(2)}(b) = \frac{1}{12}h^2 \quad \text{and} \quad g^{(3)}(a) = -g^{(3)}(b) = -\frac{1}{2}h.$$

Hence our identity becomes

$$\int_a^{a+h} f(x)\, dx = \frac{1}{2}h\left[f(a) + f(a+h)\right]$$
$$+ \frac{1}{12}h^2\left[f'(a) - f'(a+h)\right]$$
$$+ \int_a^{a+h} s_h(x-a)f^{(4)}(x)\, dx \qquad (9.7)$$

for all $f \in C^4([a, a+h])$.

The last identity leads to two different formulas for $\int_{a-h}^{a+h} f(x)\, dx$. One comes from replacing h by $2h$ and a by $a - h$:

$$\int_{a-h}^{a+h} f(x)\, dx = h\left[f(a-h) + f(a+h)\right]$$
$$+ \frac{1}{3}h^2\left[f'(a-h) - f'(a+h)\right]$$
$$+ \int_{a-h}^{a+h} s_{2h}(x-a+h)f^{(4)}(x)\, dx. \qquad (9.8)$$

Or we can use (9.7) to integrate over $[a-h, a]$ and $[a, a+h]$ separately and add the resulting equations. To simplify the addition of the resulting integrals, we unify the formulas for the integrands. On $[a - h, a]$, we have

$$s_h (x - (a - h)) = \frac{1}{24} (x - a + h)^2 (x - a)^2$$
$$= s_h (|x - a|),$$

and that agrees with $s_h (x - a)$ on $[a, a + h]$. So we find

$$\int_{a-h}^{a+h} f(x)\, dx = \frac{1}{2} h \left[f(a - h) + 2f(a) + f(a + h) \right]$$
$$+ \frac{1}{12} h^2 \left[f'(a - h) - f'(a + h) \right]$$
$$+ \int_{a-h}^{a+h} s_h (|x - a|) f^{(4)}(x)\, dx. \qquad (9.9)$$

We get a still more useful formula for $\int_{a-h}^{a+h} f(x)\, dx$ by subtracting $1/3$ of (9.8) from $4/3$ of (9.9) to make the h^2 terms cancel:

$$\int_{a-h}^{a+h} f(x)\, dx = \frac{1}{3} h \left[f(a - h) + 4f(a) + f(a + h) \right]$$
$$+ \int_{a-h}^{a+h} K_h (x - a) f^{(4)}(x)\, dx, \qquad (9.10)$$

where we've called

$$K_h (x) = \frac{4}{3} s_h (|x|) - \frac{1}{3} s_{2h} (x + h).$$

Equation (9.10) is the formula we need to analyze Simpson's rule.

The first term on the right of equation (9.10) is Simpson's approximation to $\int_{a-h}^{a+h} f(x)\, dx$. The second term is the error in the approximation when f is in $C^4 [a - h, a + h]$. Obviously there is no error at all when f is a polynomial of degree three or less. To understand the error in general, we need to examine $K_h (x - a)$ on the interval $[a - h, a + h]$, and we might as well take $a = 0$. Since

$$s_{2h} (x + h) = \frac{1}{24} (x + h)^2 (x - h)^2 = \frac{1}{24} (x^2 - h^2)^2,$$

we see that $K_h (x)$ is an even function, and for $x \geq 0$ we have

$$K_h (x) = \frac{4}{72} x^2 (x - h)^2 - \frac{1}{72} (x + h)^2 (x - h)^2$$
$$= \frac{1}{72} (x - h)^2 \left[4x^2 - (x + h)^2 \right].$$

Since

$$4x^2 - (x+h)^2 = (3x+h)(x-h) = 3(x-h)^2 + 4h(x-h),$$

we see that $K_h(x) \le 0$ on $[0, h]$ with

$$\int_0^h K_h(x)\, dx = \frac{1}{72} \int_0^h \left[3(x-h)^4 + 4h(x-h)^3 \right] dx$$

$$= \frac{1}{72} \left[\frac{3}{5} h^5 - h^5 \right] = -\frac{1}{180} h^5.$$

By symmetry, $K_h(x) \le 0$ on $[-h, h]$ as well, and

$$\int_{-h}^h K_h(x)\, dx = -\frac{1}{90} h^5.$$

Since $K_h(x-a)$ doesn't change sign on $[a-h, a+h]$, when $f^{(4)}$ is continuous there the mean value theorem for integrals lets us express the error conveniently as

$$\int_{a-h}^{a+h} K_h(x-a) f^{(4)}(x)\, dx = -\frac{1}{90} h^5 f^{(4)}(\xi)$$

for some ξ in the interval $[a-h, a+h]$.

Since the error in using Simpson's approximation for integration over the interval $[a-h, a+h]$ seems to be on the order of h^5, we don't want to use it directly for integration over long intervals. A much better procedure is to subdivide an interval $[a, b]$ into shorter subintervals, use Simpson's approximation on each, and then add the results. The easiest thing to do is to use equal subintervals; just pick an integer n and call

$$h = \frac{1}{2n}(b-a) \quad \text{and} \quad x_k = a + kh \quad \text{for } k = 0, 1, \ldots, 2n.$$

Since $[x_{2k-2}, x_{2k}] = [x_{2k-1} - h, x_{2k-1} + h]$, we have

$$\int_a^b f(x)\, dx = \sum_{k=1}^n \int_{x_{2k-2}}^{x_{2k}} f(x)\, dx$$

$$= \sum_{k=1}^n \frac{1}{3} h \left[f(x_{2k-2}) + 4f(x_{2k-1}) + f(x_{2k}) \right]$$

$$- \frac{1}{90} \sum_{k=1}^n h^5 f^{(4)}(\xi_k).$$

The first of these last two sums can be regrouped as the form of Simpson's approximation most commonly used:

$$\frac{1}{3}h\left[f\left(x_0\right) + 4f\left(x_1\right) + 2f\left(x_2\right) + 4f\left(x_3\right) + 2f\left(x_4\right) + \cdots\right.$$
$$\left.\cdots + 2f\left(x_{2n-2}\right) + 4f\left(x_{2n-1}\right) + f\left(x_{2n}\right)\right].$$

The second one represents the error; let's find a simpler form for it. We've assumed $f^{(4)}$ to be continuous on $[a, b]$, so the intermediate value theorem guarantees that the average of any finite set of its values on $[a, b]$ is another value. Hence the error is

$$-\sum_{k=0}^{n-1}\frac{1}{90}h^5 f^{(4)}\left(\xi_k\right) = -\frac{n}{90}h^5 f^{(4)}\left(\xi\right) = -\frac{b-a}{180}h^4 f^{(4)}\left(\xi\right)$$

for some $\xi \in [a, b]$. That's the usual representation of the error in Simpson's rule.

EXERCISES

7. Check the last formula for the error in Simpson's rule by comparing the exact value of $\int_0^{2h} x^4\,dx$ to Simpson's approximation.

8. Simpson's rule can still give a useful approximation when the function to be integrated isn't in C^4, but our representation for the error is no longer valid. Show that when Simpson's rule is used to approximate $\int_0^1 \sqrt{x}\,dx$, the error is always at least $\left(\sqrt{2} - \frac{4}{3}\right) h^{3/2}$. Examining $\int_0^{2h} \sqrt{x}\,dx$ and $\int_{2h}^1 \sqrt{x}\,dx$ separately can help since $\sqrt{x} \in C^4\,[2h, 1]$.

9. Approximate $\log x$ for $x > 1$ by using Simpson's rule with $h = \frac{1}{2}\left(x - 1\right)$ to approximate the integral defining it, and find upper and lower bounds for the error in this approximation when $1 < x \le 3$.

10. Use the approximation from the previous problem and Newton's method to approximate e, the solution of $\log x = 1$. Find an algebraic equation whose solution should give a better estimate.

4   THE SUBSTITUTION RULE FOR INTEGRALS

Typically, students in calculus first encounter substitutions as an aid to finding antiderivatives:

$$\int f\left(g\left(x\right)\right) g'\left(x\right)\,dx = \int f\left(u\right)\,du \quad \text{with } u = g\left(x\right).$$

This is simply a help in keeping track of the effect of the chain rule. When we identify a function F with $F'\left(u\right) = f\left(u\right)$, we can then appeal to the fundamental theorem of calculus:

$$\int_a^b f\left(g\left(x\right)\right) g'\left(x\right) \, dx = F\left(g\left(x\right)\right)\big|_a^b = F\left(g\left(b\right)\right) - F\left(g\left(a\right)\right)$$

The last quantity is recognized as $\int_{g(a)}^{g(b)} f\left(u\right) \, du$, and that's the origin of the substitution rule for integrals. But there are some potential difficulties that this approach ignores, and they can come back to haunt us. For example, when we start wondering whether $\int_a^b f\left(g\left(x\right)\right) g'\left(x\right) \, dx$ and $\int_{g(a)}^{g(b)} f\left(u\right) \, du$ are equally amenable to numerical approximation, we discover that there can be some real surprises.

Here's an example involving a simple but unfortunate substitution. For $f\left(u\right) = u^{-1/4}$ and $g\left(x\right) = x^{4/3}$, we see easily that

$$\int_0^8 f\left(g\left(x\right)\right) g'\left(x\right) \, dx = \int_0^8 \left(x^{4/3}\right)^{-1/4} \cdot \frac{4}{3} x^{1/3} dx$$
$$= \frac{4}{3} \int_0^8 dx = \frac{32}{3}.$$

But

$$\int_{g(0)}^{g(8)} f\left(u\right) \, du = \int_0^{16} u^{-1/4} \, du,$$

and $u^{-1/4}$ is not Riemann integrable over $[0, 16]$. Trying to apply Simpson's rule to $\int_0^{16} u^{-1/4} \, du$ quickly reminds us why such integrals are called improper and why that distinction is important.

Now that we're aware that integrability may be a problem, we're ready to state and prove a simple version of the substitution theorem. We'll prove it by using Theorem 2.1 in Chapter 5.

THEOREM 4.1: *Let g be a continuous, increasing function on $[a, b]$, with g differentiable everywhere in (a, b) and g' Riemann integrable over $[a, b]$. Suppose also that f is Riemann integrable over $[g\left(a\right), g\left(b\right)]$. Then $f\left(g\left(x\right)\right) g'\left(x\right)$ defines a Riemann integrable function over $[a, b]$, and*

$$\int_a^b f\left(g\left(x\right)\right) g'\left(x\right) \, dx = \int_{g(a)}^{g(b)} f\left(u\right) \, du.$$

☐ *Proof:* We'll need to partition $[a, b]$ and $[g\left(a\right), g\left(b\right)]$ simultaneously, and the properties of g make this easy. For $\mathcal{P} = \{x_0, x_1, \ldots, x_n\}$ a partition of $[a, b]$, we'll call $\mathcal{P}' = \{u_0, u_1, \ldots, u_n\}$ the partition of

$[g(a), g(b)]$ with $u_k = g(x_k)$ for $k = 0, 1, 2, \ldots, n$. Since g is continuous and monotonic on $[a, b]$, every partition of $[g(a), g(b)]$ can be expressed as $\mathcal{P}'$ for some partition $\mathcal{P}$ of $[a, b]$. Given arbitrary sampling points $\{x_k^*\}_{k=1}^n$ for $\mathcal{P}$, we call $u_k^* = g(x_k^*)$. Since g is monotonic the set $\{u_k^*\}_{k=1}^n$ will be a set of sampling points for $\mathcal{P}'$.

The key to the proof is the mean value theorem; there is at least one point ξ_k in each interval (x_{k-1}, x_k) such that

$$\Delta u_k = g(x_k) - g(x_{k-1}) = g'(\xi_k) \Delta x_k.$$

Consequently,

$$f(g(x_k^*)) g'(x_k^*) \Delta x_k - f(u_k^*) \Delta u_k = f(u_k^*) \left[g'(x_k^*) - g'(\xi_k) \right] \Delta x_k$$

for each k. Transposing the second term and summing from $k = 1$ to n, we obtain

$$\sum_{k=1}^n f(g(x_k^*)) g'(x_k^*) \Delta x_k = \sum_{k=1}^n f(u_k^*) \left[g'(x_k^*) - g'(\xi_k) \right] \Delta x_k$$

$$+ \sum_{k=1}^n f(u_k^*) \Delta u_k. \tag{9.11}$$

Equation (9.11) points the way to proving the substitution theorem since the sum on the left represents an arbitrary Riemann sum for $(f \circ g) g'$ associated with $\mathcal{P}$ and the last sum on the right is a Riemann sum for f associated with $\mathcal{P}'$. Our strategy is to show that for any $\varepsilon > 0$, by choosing $\mathcal{P}$ carefully we can guarantee that the first sum on the right of (9.11) is smaller than $\frac{1}{2}\varepsilon$ and that the last sum is within $\frac{1}{2}\varepsilon$ of $\int_{g(a)}^{g(b)} f(u)\, du$.

Since f is Riemann integrable over $[g(a), g(b)]$ by hypothesis, when $\varepsilon > 0$ is given we know there is a partition $\mathcal{P}_0'$ of $[g(a), g(b)]$ with

$$\mathcal{R}(\mathcal{P}_0') \subset \left(\int_{g(a)}^{g(b)} f(u)\, du - \frac{1}{2}\varepsilon, \int_{g(a)}^{g(b)} f(u)\, du + \frac{1}{2}\varepsilon \right).$$

Furthermore, every Riemann sum associated with any refinement of $\mathcal{P}_0'$ must be in the same interval. The partition $\mathcal{P}_0'$ corresponds to a partition $\mathcal{P}_0$ of $[a, b]$, and choosing $\mathcal{P}$ to be a refinement of $\mathcal{P}_0$ will make $\mathcal{P}'$ be a refinement of $\mathcal{P}_0'$.

To find out how to choose $\mathcal{P}$, we call m_k and M_k the infimum and supremum of the values of g' over the subinterval $[x_{k-1}, x_k]$, and choose a number M with $|f(u)| \leq M$ for all $u \in [g(a), g(b)]$. Then we have

$$\left| f(u_k^*) \left[g'(x_k^*) - g'(\xi_k) \right] \Delta x_k \right| \leq M \left[M_k - m_k \right] \Delta x_k,$$

so

$$\left| \sum_{k=1}^{n} f\left(u_k^*\right) \left[g'\left(x_k^*\right) - g'\left(\xi_k\right)\right] \Delta x_k \right| \leq \sum_{k=1}^{n} M\left[M_k - m_k\right] \Delta x_k$$
$$= M\left[U\left(\mathcal{P}\right) - L\left(\mathcal{P}\right)\right],$$

where $U\left(\mathcal{P}\right)$ and $L\left(\mathcal{P}\right)$ are the upper and lower sums for $g'\left(x\right)$. Since g' is Riemann integrable by hypothesis, we can choose a refinement $\mathcal{P}$ of $\mathcal{P}_0$ to make this last expression be smaller than $\frac{1}{2}\varepsilon$, and that completes the proof.   ∎

It's not hard to extend the substitution theorem somewhat. With just a slight change in our arguments, we can also prove a version where g is decreasing instead of increasing. And since we know how to add integrals over adjacent subintervals, we can also extend it to any case where we can split $[a, b]$ into finitely many subintervals, with all the needed hypotheses satisfied on each. Accordingly, we state a more general version of the substitution theorem below; we won't provide any further arguments for its validity.

THEOREM 4.2:   *Let g be a continuous function on $[a, b]$, and suppose that $g'\left(x\right)$ exists at all but finitely many points in $[a, b]$, agreeing with a Riemann integrable function at the points where it is defined. Suppose also that there is a partition $\mathcal{P} = \{x_0, \dots, x_n\}$ of $[a, b]$ such that g is monotonic on each subinterval $[x_{k-1}, x_k]$. If f is Riemann integrable over the image of $[a, b]$ under g, then $f\left(g\left(x\right)\right) g'\left(x\right)$ is Riemann integrable over $[a, b]$ and*

$$\int_a^b f\left(g\left(x\right)\right) g'\left(x\right) \, dx = \int_{g(a)}^{g(b)} f\left(u\right) \, du.$$

The important thing to notice about this more general theorem is that simply assuming the integrability of f over $[g\left(a\right), g\left(b\right)]$ is not enough; we need integrability over the entire image of $[a, b]$. For example, consider

$$\int_a^b f\left(g\left(x\right)\right) g'\left(x\right) \, dx = \int_{-1}^{2} \left[\frac{1}{\sqrt{x^2 + 1} - 1}\right] \frac{x \, dx}{\sqrt{x^2 + 1}}.$$

Here $g\left(x\right) = \sqrt{x^2 + 1}$, so that $[g\left(a\right), g\left(b\right)] = [\sqrt{2}, \sqrt{5}]$, and $f\left(u\right) = \left(u - 1\right)^{-1}$ is Riemann integrable over that interval. However, the image of $[-1, 2]$ under g is $[1, \sqrt{5}]$, and f is not integrable over that interval. And in fact, $f\left(g\left(x\right)\right) g'\left(x\right)$ is not integrable over $[-1, 2]$; it's not even considered improperly integrable.

In summary, we must be careful when making substitutions in definite integrals. Unlike matter and energy, integrability can easily be created or destroyed.

EXERCISES

11. The substitution theorem can be proved with slightly different assumptions. Instead of assuming that g is continuous on $[a, b]$ and differentiable on (a, b) with its derivative g' Riemann integrable, we can assume that g' is a given Riemann integrable function and that g satisfies

$$g(x) = g(a) + \int_a^b g'(t) \, dt \quad \text{for } a \le x \le b.$$

Assumptions about the monotonicity of g are then conveniently given in terms of the sign of g'. Assume that $g' > 0$ on $[a, b]$ and prove the substitution theorem with these modified hypotheses.

12. Show that if f is Riemann integrable over $[-1, 1]$, then

$$\int_a^b f(\sin x) \cos x \, dx = \int_{\sin a}^{\sin b} f(u) \, du \quad \text{for every } a, b \in \mathbf{R}.$$

13. What assumptions are needed to obtain

$$\int_a^b f(x) \, dx = \int_{\arcsin(a)}^{\arcsin(b)} f(\sin t) \cos t \, dt?$$

14. Suppose that $g'(x)$ is positive and continuous on $[a, b]$ and that f is not Riemann integrable over the entire interval $[g(a), g(b)]$ but only over all subintervals $[g(t), g(b)]$ with $t \in (a, b)$. Prove that

$$\lim_{t \to a+} \int_t^b f(g(x)) g'(x) \, dx = \lim_{s \to g(a)+} \int_s^{g(b)} f(u) \, du$$

in the sense that if either limit exists, then so does the other and the limits are equal.

REFERENCES

1. R. P. Boas, *A Primer of Real Functions*, 3rd ed., Providence, RI: Mathematical Association of America, 1994.
2. E. Landau, *Foundations of Analysis*, New York: Chelsea Publishing Co., 1951.
3. G. F. Simmons, *Calculus with Analytic Geometry*, New York: McGraw-Hill Book Company, 1985.
4. C. B. Boyer, *A History of Mathematics*, New York: John Wiley & Sons, 1968.
5. W. Rudin, *Principles of Mathematical Analysis*, 3rd ed., New York: McGraw-Hill Book Company, 1976.

INDEX